Contractor Safety Management

Contractor safety management represents one of the most critical yet overlooked challenges facing organizations today. When accidents occur involving contractors and subcontractors, the spotlight increasingly falls on client organizations who face mounting reputational, legal, and financial consequences. *Contractor Safety Management: From Compliance to Culture* examines how responsibility for safety is shared across complex contractor relationships, which moves beyond the traditional approach of treating safety as a compliance checkbox exercise.

Drawing on 17 major case studies from around the world, including the Grenfell Tower fire, the Florida International University pedestrian bridge collapse, and various global energy projects, this book reveals how cultural differences, procurement practices, and leadership decisions directly impact safety outcomes. Written in an engaging, narrative-driven style, it combines real-world lessons with practical frameworks and original methodologies that readers can implement immediately. The book provides ready-to-use tools for strengthening contractor oversight, risk management, and accountability and demonstrates how to embed safety into contracts from the beginning, ultimately protecting organizations from devastating consequences.

This book is for safety professionals, business leaders, senior executives, frontline safety managers, and policymakers who work with contractors and subcontractors. Its appeal will extend to organizations operating in multinational environments across industries including construction, oil and gas, retail, and infrastructure, providing the cultural intelligence and practical guidance needed to build resilient contractor safety systems that protect both people and business interests.

Contractor Safety Management
From Compliance to Culture

Adel Lawson

CRC Press
Taylor & Francis Group
Boca Raton London New York

CRC Press is an imprint of the
Taylor & Francis Group, an **informa** business

First edition published 2026
by CRC Press
2385 NW Executive Center Drive, Suite 320, Boca Raton FL 33431

and by CRC Press
4 Park Square, Milton Park, Abingdon, Oxon, OX14 4RN

CRC Press is an imprint of Taylor & Francis Group, LLC

ISBN: 978-1-041-27213-7 (hbk)
ISBN: 978-1-041-27383-7 (pbk)
ISBN: 978-1-003-75412-1 (ebk)

DOI: 10.1201/9781003754121

Typeset in Times
by KnowledgeWorks Global Ltd.

Contents

About the Author

Adel Lawson is an Occupational Health, Safety, Environmental and Risk Management professional and Managing Director of Demoura Lawson Consulting. With more than two decades of experience spanning corporate enterprise, oil and gas, marine operations, aviation, large-scale construction, infrastructure, and manufacturing environments, he has built a career at the intersection of operational performance, regulatory compliance, and strategic risk governance.

Throughout his career, Adel has led the development of integrated management systems aligned with international standards, helping organizations achieve and maintain ISO (International Organization for Standardization) certifications across quality, health and safety, and environmental disciplines. His work extends beyond documentation and compliance frameworks; he is known for embedding practical, operationally effective safety cultures within complex, contractor-heavy environments where performance, time pressure, and risk coexist.

As a chartered fellow of the Institute of Safety and Health (IOSH), he has contributed to nationally significant programmes and high-profile infrastructure developments, advising on occupational safety governance, contractor oversight, and regulatory compliance systems. His expertise includes enterprise risk alignment, safety leadership strategy, workplace inspection regimes, major project safety assurance, and the implementation of behavioral and systems-based safety models.

Mr. Lawson's approach combines engineering insight with behavioral understanding. Trained initially as a mechanical engineer, he later specialized in occupational safety and health, enabling him to bridge the technical, human, and managerial dimensions of risk. His academic background includes a Bachelor of Science in Mechanical Engineering from Northumbria

University, UK, and a Master of Science in Occupational Safety and Health from Columbia Southern University, USA.

Beyond compliance, his work focuses on resilience, designing systems that withstand operational strain, contractor interface risk, and cultural complexity. He has advised organizations on improving audit performance, strengthening governance structures, enhancing safety reporting frameworks, and integrating environmental and sustainability principles into core business strategy.

Recognized for his pragmatic yet strategic perspective, Mr. Lawson advocates for safety as a leadership discipline rather than a regulatory obligation. His work reflects a consistent belief that effective health, safety, and environmental management is not an administrative exercise but a fundamental driver of organizational excellence and long-term value.

Preface

Contractor Safety Management: From Compliance to Culture isn't a checklist manual. It's a practical guide for organizations that take their duty of care seriously … especially in a world where more work than ever is done by people outside the organization.

Over eight chapters, this book explores the shift from internal employee safety to contractor safety, showing why the old mindset of "we've hired them, so it's their problem" no longer holds water, legally, ethically, or reputationally. Starting with the changing global workforce, the book makes the case for why strong contractor management isn't just good practice, it's essential business strategy.

Through real-world examples, from wind farms to road markings, from collapsed buildings to offshore disasters, we look at what happens when clients fail to manage contractor safety, and how things can go right when they do. Topics include pre-contract planning, choosing the right contract type, embedding safety key performance indicators (KPIs), effective monitoring, and the often-overlooked importance of cultural awareness in international teams.

This book doesn't promise quick fixes. It offers grounded, experience-based insight for safety professionals, project managers, procurement officers, and senior leaders who understand that reputation, performance, and lives are all on the line when contractors are on the ground.

It's a call to stop treating contractor safety as an afterthought and to start managing it like the critical, strategic function it really is.

Acknowledgments

Writing Contractor Safety Management: From Compliance to Culture has been a journey shaped by the people who supported, encouraged, and inspired me long before the first chapter was drafted. My deepest gratitude goes to my family, Carina, Mila, Nina, my mother, and my father, whose patience and unwavering belief carried me through every round of revisions, and every moment spent listening to loud audio readings of the manuscript. Their love, humor, and grounding presence have been my anchor throughout this process.

A special thanks goes to my dad, whose thoughtful feedback and help during the editing stages strengthened this book in countless ways.

I am equally grateful to my colleagues, mentors, and the many professionals I have worked with across Qatar, the Middle East, and beyond. Past and present, each of you has contributed to my understanding of what safety leadership truly means. Your stories, insights, and lived experiences helped shape the heart of this book.

To the organizations and teams who trusted me in their operations, thank you for allowing me to learn alongside you. This book is as much yours as it is mine.

The Changing World 1

Every age has its own way of doing things, and they seem at the time to be the right, natural or only way things should be done. It's myopia! Eighty years ago, the poet Louis MacNeice wrote that, "World is crazier and more of it than we think, incorrigibly plural." There's a simple way to rid yourself of the error of thinking, "This is how we do it, so this must be the right way." Just look at what you're doing—almost anything you're doing—and ask yourself, "What would my grandparents make of this? And what about their grandparents?" And yet, those ancients were as convinced as we are that what they were doing must be correct.

At least two things that seem natural to us living in the 2020s would have amazed anyone living as recently as the 1950s (and many people who lived in the 1950s are still living today). The first is the move to a gig economy. The second is globalization. And the second tells us something about the first, because one thing globalization has done is to remove from people in developed countries—by which I mean what we generally refer to as the West—protections never enjoyed in the rest of the world.

By gig economy, we mean a labor market where short-term contracts or freelance work rather than permanent jobs are common. It is the commodification of skills on-demand. Workers float between projects, sometimes for different companies, with little job security but with flexibility and, for some, a sense of autonomy. Enabled by globalization and driven by digital platforms, the gig economy is the world's new normal.

In the West in the 1950s, the norm was long-term employment. The majority of work done by most companies was performed by employees. And that was still true in the 1960s, the 1970s and, though its hold was weakening, the 1980s. It isn't true any longer. Today, even in the West, more of the operational work is probably done by contractors than by employees. The West has become a freelance world. A gig economy.

Why do I keep stressing the West? Because permanent employment wasn't true elsewhere in the world. In emerging economies, most people picked up work where and when they could. By the 1990s, people in the West had started to talk about "portfolio work"—workers were encouraged

DOI: 10.1201/9781003754121-1

to develop a range of skills that could help them find employment, and the employment would mostly be freelance or short-time contract work. In the 30 years since portfolio working was first discussed, freelancing has become the norm everywhere—companies that want to be profitable have as few permanent employees as they can and hire others to do what needs to be done at any given time. The reason for the change is globalization, and the engine that has fired it is the internet.

But this book is not about portfolio working. This book is about safety, and who handles it. A gig economy is more efficient and more cost-effective—what it is not is responsibility-free. Traditionally, it was the employer who was responsible for worker safety. If you were a British, American, or European coal miner in the 19th century, you might have found evidence for that thin on the ground; nevertheless, let's take it as our starting point. So, if the person responsible for a worker's safety is the worker's employer, who takes that responsibility for a freelance contractor? The gang master? The agency through whom the contractor is hired? The contractor him or herself? The premise on which this book is written, and which I ask you to accept, is that all of these actors are responsible for the contractor's safety, but so—and possibly above all—is the client for whom the contractor is working.

Responsibility is legal and not simply ethical. Around the world, the law has changed—or is changing—to place ultimate responsibility for the safety and well-being of everyone involved in a project on the project's true owner or client. A growing body of regulation makes this explicit. In the Middle East, Qatar's **Law No. 14 on Labor** outlines the employer's duty to provide a safe working environment and allows for direct penalties on organizations that outsource risk without oversight. In the UK, the **Construction Design and Management Regulations (CDM)** put clear legal duties on clients, designers, and contractors to plan and manage risks from day one. In Australia, the **Work Health and Safety (WHS) Act** shifts the conversation further, defining a "Person Conducting a Business or Undertaking" (PCBU) and making them legally responsible—even if the risk is created by someone further down the contracting chain.

This is a global trend. **Occupational Safety and Health Administration (OSHA)** regulations in the United States have consistently reinforced that general contractors and host employers can be held accountable for the safety of subcontractors and temporary workers. In Canada, provincial laws—such as Ontario's **Occupational Health and Safety Act**—impose clear obligations on project owners and constructors to ensure health and safety across worksites.

Further afield, South Africa's **Construction Regulations under the Occupational Health and Safety Act** require a client to not only appoint a competent contractor but also ensure a comprehensive health and safety specification is part of the tendering process. In Asia, Singapore's **Workplace Safety and Health Act** holds companies accountable for ensuring every person affected by the work—employee or not—is kept safe. Even in jurisdictions where enforcement is still catching up, the legislative direction is clear: clients can no longer delegate accountability when something goes wrong.

Failure to act on this responsibility is no longer simply a moral failing—it's criminal. Courts across the globe are increasingly prepared to impose fines, sanctions, and even custodial sentences where preventable harm is caused due to a lack of client oversight. Ignorance is no defense, and silence in a contract is no shield.

There is no shortage of examples showing how catastrophic the consequences can be when contractor oversight is lax. In 1978, in Willow Island, West Virginia, 51 workers fell to their deaths when scaffolding collapsed on a cooling tower under construction. The workers, employed by contracted firms, were operating under severe time constraints and lacked the protective safeguards that should have been mandated by a robust contractor safety management system. Investigations revealed the scaffolding had been affixed to concrete that had not properly cured, and substandard materials were used—choices driven by schedule pressures and inadequate oversight. No clear communication or enforced stop-work authority existed between the client and contracting tiers. This was not merely an engineering failure but a stark example of how tragic consequences can result from inadequate integration of contractor safety into the client's management system. It reinforces the point: unless contractor safety is treated with equal priority to internal operations, lives are put at unnecessary risk.

Or take Singapore in 2004. The Nicoll Highway collapse, during Mass Rapid Transit (MRT) construction, killed four workers and halted the project for months. Improper retaining wall design and a failure to monitor ground movement adequately were identified as causes—but, at heart, it was a failure of oversight. Contractors weren't properly integrated into the host organization's risk management systems.

In Philadelphia in 2013, a building being demolished collapsed onto a thrift store, killing six people. The contractor had no permit to remove load-bearing walls, and there was no adequate supervision. It was a case of neglect, and it led to criminal charges.

Australia has not been immune. In 2006, in Melbourne, a subcontractor's worker fell through a roof with no safety mesh in place. ACR Roofing, the principal contractor, was found liable. The legal system sent a clear message: even when you subcontract, responsibility remains yours.

Sometimes, the consequences reach beyond the workplace. In Brooklyn, New York, 2019, a stone fence collapsed, killing a five-year-old child. The contractor hadn't filed for the correct permits or followed engineering guidance. He was found guilty of negligent homicide. For the client organization, the fallout was severe. Public outrage did not distinguish between the contractor and the property owner. Lawsuits were filed against the property developer for failing to ensure due diligence in contractor selection and oversight. The reputational damage affected investor confidence, and the site was shut down pending investigation. This tragedy serves as a stark reminder that inadequate contractor safety management can have direct legal, financial, and reputational consequences for client organizations—even when the contractor appears to be solely at fault.

In Perth in 2022, a 12-year-old boy gained access to an unsecured construction site and was killed by a falling beam. PTG Construction was fined for failing to secure the site. A simple barrier could have prevented that death.

In Sydney, Scopeview Projects ignored repeated safety warnings before a wall collapsed and severely injured a worker in 2020. Despite citations, no changes were made. The court found both the company and its director criminally liable.

In Glasgow, Scotland that same year, a telehandler operator died after his vehicle slipped down an embankment. Merchant Homes was fined £160,000. No risk assessment had been done. No-one asked: is this terrain safe for heavy machinery? The client organization not only bore the weight of the fine but also suffered reputational damage, particularly in the local community and the construction sector. Public scrutiny focused not just on the incident, but on why the client had not insisted on a more rigorous contractor safety protocol or ensured that a basic terrain and vehicle risk analysis was in place. Insurance premiums rose, tenders were questioned, and internal reviews were launched. It was a sobering lesson that oversight of contractor safety is not just a duty that can be delegated to others.

In 2025, India witnessed the death of an 18-year-old laborer when part of a hillock collapsed during highway construction. The contractor had ignored basic geotechnical assessments. In Gurgaon the same year, water accumulation caused a basement wall to collapse, killing one worker and injuring others. Soil erosion had been evident for days. No action had been taken.

Each of these cases speaks to a wider truth: that when things go wrong, the public does not care whose payroll the dead were on. The public, the media, the courts all hold the project owner accountable.

As seen during preparations for the Qatar 2022 FIFA World Cup, scrutiny extended not just to the country but to every contractor and subcontractor involved. The infamous "6,500 migrant workers have died in Qatar" headline created an enduring image that governments and corporations alike have struggled to refute. While statistics were misused or misinterpreted, the reputational damage had already been done. This public pressure prompted significant labor reforms in Qatar: wage protection systems, mandatory heat stress legislation, and the dismantling of the Kafala system which had given employers of migrant workers almost total control over migrant workers' employment and immigration status. Qatar's actions reflect the shifting tide: reputational risk and moral obligation are increasingly intertwined.

It's a question of culture. One thing that globalization has not changed is the importance of cultural differences, particularly in respect of safety. You might find elaborate personal protective equipment (PPE) and lane closures on British highways, while a road-marking crew in Kenya works inches away from speeding traffic with little more than hope and a flag. In China, where workforce abundance has historically devalued individual lives, safety culture has lagged … though this too is changing.

If you think wellness is a Western luxury, think again. This book will delve deeply into worker welfare, emphasizing its integral role in contractor safety considerations. As the contractor workforce becomes increasingly globalized, welfare is no longer a peripheral concern, it has become central. In the Middle East, many migrant worker suicides were linked to social isolation, lack of cultural support, and harsh living conditions. Wellness, in contractor safety includes mental health, hydration, nutrition, and humane accommodation. These are not optional extras, they're fundamental to productivity, dignity, and compliance. The scandal involving BYD in Brazil is a case in point. In late 2024, Brazilian authorities suspended construction on BYD's electric vehicle factory after over 160 Chinese migrant workers were found living in overcrowded, unsanitary conditions with their passports confiscated and wages withheld. These individuals had been hired by a subcontractor, Jinjiang Construction Brazil, but public outrage was squarely directed at BYD. The reputational damage was swift and global, highlighting how lapses in contractor welfare management now carry the greatest reputational consequences for the client. This incident has shown how reputational risk from worker welfare failures can eclipse even those stemming from physical safety failures, and it underscores why welfare is now deeply interwoven into the broader fabric of contractor safety.

Well-being failures are punished not just legally but socially and economically. Cancel culture now reaches into boardrooms. Reputational damage can crash share prices and end careers. Social media exposes in seconds what regulators might take months to confirm.

There was a time when an incident involving a subcontractor might stay buried in an internal report or fade quietly into a safety committee's minutes. That time is over. In today's hyperconnected world, contractor failures don't just leave scars on your site, they go viral and the impact on your reputation is global and permanent.

In 2019, **ProPublica** and **BuzzFeed News** published investigations into Amazon's delivery network, revealing how the company's use of third-party delivery contractors led to fatal road accidents. Amazon did not directly employ these drivers, but that was a distinction not made by the public. Photos of unsafe delivery practices, driver fatigue, and crashes began circulating on Twitter, Reddit, and Instagram. The visual evidence told a damning story.

Amazon's stock price wasn't drastically impacted, but its reputation as a safe and ethical employer took a hit, particularly among younger, socially conscious consumers. The online conversation forced the company to reexamine contractor policies, increase vetting, and implement new accountability protocols. This wasn't prompted by regulation; it was driven by perception, and perception was driven by social media.

McDonald's has long operated under a franchise model, but when contractor-managed labor practices within certain locations were found to be abusive, including wage theft and unsafe conditions, activists took to TikTok and YouTube. Videos from employees went viral, showing real-time violations. The phrase "McBroken" trended not because of ice cream machines, but because of broken promises on workplace safety and fairness. Again, the legal employer might have been the franchisee, but the public held McDonald's accountable.

Social media doesn't care about contractual distinctions. If a contractor's behavior reflects poorly on your brand, *you* take the blame. In a digital age, contractor management isn't just a safety issue, it's a reputational safeguard. Because now, the court of public opinion meets in comment sections, and its judgments are swift, emotional, broadcast globally, and difficult to reverse.

The following story is well known among both occupational safety and investment professionals. In 1987, Paul O'Neill had been appointed CEO of Alcoa (Aluminum Company of America). He addressed an investors' meeting that had finance professionals who were there in a flat spin. They arrived expecting a presentation on all the things O'Neill was going to do to cut costs, boost sales, and increase the company's net worth. They needed to hear that—or thought they did—because the company had not been doing well and they wanted to know what he intended to do about it. In fact, his opening words were: "I want to talk to you about worker safety. Every year, numerous Alcoa workers are so badly injured they miss work. I intend to make Alcoa the safest company in America. I intend to go for zero injuries."

When his listeners asked questions that were about profitability and P/E ratios rather than worker safety, O'Neill said, "I'm not certain you heard me. If you want to understand how Alcoa is doing, look at our workplace safety figures."

That meeting, and the battle that followed it to change the Alcoa culture, have been widely covered elsewhere and there's no need for me to go into detail here. All I'll say is that, when O'Neill stepped down 13 years later, Alcoa's income was five times higher than it had been when he became CEO.

Those investors had got what they wanted, too, because the market valuation was up from $3 billion to more than $27 billion. The costs incurred in building a safety culture do not need to be money spent for no reward. The opposite can be true.

O'Neill's genius wasn't in ignoring financial outcomes—it was in understanding that *health and safety could be the foundation* for operational excellence. He saw safety not as a compliance obligation but as a **keystone habit**. Improve safety, and you improve everything.

What is a keystone habit? It's a behavior that, when changed, has a ripple effect across an organization. Safety, when properly understood and authentically prioritized, influences quality, communication, trust, decision-making, and even innovation. Because when people see that leadership genuinely cares about their well-being, they begin to care more about their work, their teammates, and the company's mission. That's culture—and culture is scalable.

Let's break that down with clarity: building a strong safety culture demands clear communication, efficient systems, engaged supervision, and proactive feedback loops. These aren't just safety tools—they're *business tools.* And once in place, they rarely stay within the confines of safety. They spill over into production processes, customer service, and organizational learning.

Consider research published in the *Harvard Business Review* that found a strong correlation between high-trust workplace cultures and improved performance across multiple indicators—reduced turnover, higher profitability, greater innovation, and increased employee engagement. Trust is built when people believe leadership has their best interests at heart. Nothing builds—or destroys—that trust faster than safety. A leader who cuts corners on safety signals that people are expendable. A leader who champions it, as O'Neill did, sends the opposite message: "You matter."

Other forward-thinking CEOs have followed the same logic. When Satya Nadella took over as CEO of Microsoft, he emphasized empathy, emotional intelligence, and psychological safety as cornerstones of innovation. While not framed as traditional health and safety, this focus on *human-centered performance* achieved the same effect—revitalizing a stagnant business and making it one of the world's most valuable companies.

In the industrial sector, Chevron has embedded safety into its performance management system so thoroughly that leading safety behaviors are evaluated as part of executive and frontline assessments. They've understood that *you get what you measure*, and when safety becomes an expected leadership behavior—not a bolt-on—it changes how people operate.

And this isn't just a Western concept. Emirates Global Aluminium (EGA), one of the largest industrial companies in the UAE, has driven down its Total Recordable Injury Frequency Rate (TRIFR) below many of its global peers by implementing cross-functional safety leadership teams and publicly recognizing safe behavior. Their executives talk about safety not just in audits, but in town halls, investor briefings, and boardrooms.

The message is simple: smart organizations use safety as a lens through which they understand operational discipline. Safety is not a side dish. It's the test of how well your company performs under pressure—and the smarter CEOs know it. The cost of embedding safety well? Often marginal. The return? Cultural cohesion, operational reliability, stakeholder confidence, and yes, profitability.

In short, safety done right is not an expense. It's an investment—and possibly your best one.

So what must you do?

The purpose of this book is not just to raise awareness about contractor safety—it's to build understanding. Real understanding. The kind that sticks with you when contracts are being drafted, not just when something goes wrong.

As you'll see in the chapters ahead, many of the biggest problems organizations face don't suddenly appear during project execution. They're baked in early—right at the start, in the way the contract is structured, how responsibilities are divided, and whether safety and risk management are treated as priorities or afterthoughts.

That's the standard that must be reached. Not everyone is there yet. It's not uncommon to find contracts focused on deliverables, timelines, and cost, with safety clauses bolted on at the end—if they're there at all. The assumption? That the contractor will "handle it." But that assumption is where many failures begin. When a contractor is left to interpret what "safety" means on their own terms—without clear expectations, without aligned goals, without the right checks and balances—it's only a matter of time before cracks start to show.

Take, for example, the collapse of the pedestrian bridge at Florida International University in 2018. Investigations revealed that concerns over structural cracks were known in the days leading up to the collapse, but the parties involved—multiple contractors, engineering consultants, and state agencies—lacked a clear framework for accountability and response. The result: six people died, and a university that had aimed to showcase innovation ended up under federal investigation. This wasn't just a failure of engineering. It was a failure of communication, contract clarity, and shared responsibility.

The lesson here—and in many similar incidents—is that contracts must do more than spell out what is to be built. They must define how it's to be built, by whom, under what conditions, and with what safeguards. And those safeguards aren't just about rules—they're about culture.

That's why this book explores more than compliance checklists. We'll look at contract styles—like lump sum versus time-and-materials—and how they influence safety behaviors. We'll break down what to include in your scopes of work, how to build in behavioral expectations, and how to onboard contractors in ways that actually shape outcomes on the ground. We'll also examine how shared audits, aligned reporting structures, and early risk reviews can shift safety from a reactive process to a proactive habit.

Because when it's done well, contractor safety isn't just "managed"—it becomes part of a project's DNA. It becomes a leadership issue, a strategic issue and, yes, a human issue.

So if you're reading this as a project manager, Health, Safety, and Environment (HSE) professional, procurement officer, or even a company executive, understand this: the quality and clarity of your contract—and how it's implemented—will do more to shape safety outcomes than any toolbox talk or policy memo ever will.

Forget the wellness fads. You don't need lavender-scented nap pods, smoothie bars, or Instagram-friendly yoga corners. What you need is something far less glamorous—and far more impactful. You need to **focus on the basics of human dignity**. Offer safe living conditions. Provide hydration stations. Ensure food quality. Listen. Include contractors in your safety committees. Build infrastructure for community.

This isn't just the right thing to do—it's the **smartest risk management decision you can make**. Because poor welfare provisions don't just harm the individuals involved. They fracture morale, erode performance, increase absenteeism, and open the floodgates to regulatory and reputational exposure.

Take accommodation. A contractor who sleeps in a poorly ventilated shipping container with seven other people on a mattress infested with bed bugs will not arrive on site physically or mentally prepared to carry out safe work. Yet this continues to happen across industries, especially in high-growth regions where labor demand exceeds infrastructure. Contractors may be grateful for the work—but their bodies and minds will not forget the conditions in which they live. Fatigue, dehydration, and illness are not personal weaknesses. They are predictable outcomes of neglect.

You want a well-performing site? Start with a well-rested, well-fed, and well-housed workforce.

Empathetic contractor management also means inclusion. Too often, welfare conversations are reserved for direct employees. Contractors are "temporary," "outsourced," or "not our people." But that mindset is not only outdated—it is dangerous. A person operating machinery, climbing scaffolding, or wiring electrical systems is either safe or not safe. Their employment status doesn't change the consequence of an error. That's why **leading**

organizations break this false separation. They bring contractors into their safety dialogues, into their toolbox talks, into their risk assessments, and into their HSE committees. They understand that inclusion drives ownership—and ownership drives safety.

We will talk about standards in our upcoming chapters. **ISO 45001**, the global occupational health and safety management system standard, was revolutionary not because it created new rules but because it **clarified responsibilities**. Clause 8.1.4 of ISO 45001 is explicit: organizations must coordinate with contractors on safety matters, not just dictate requirements. That means aligning hazard identification processes, sharing control measures, and reviewing performance collectively. But it goes beyond risk registers. ISO 45001 also recommends integrating *health surveillance*, *mental well-being*, and *worker participation* into the safety system. It pushes for the **systemic recognition of psychosocial risks**—stress, isolation, harassment, overwork. Not because these are trending topics, but because they are measurable threats to safe performance.

Other frameworks are following suit. The International Labour Organization's Convention 155 and Recommendation 164 also stress the equal protection of all workers—permanent, temporary, and contract. And governments are tightening the net.

But we don't need to rely on regulatory sticks to do the right thing. The **evidence of benefit** is everywhere.

Consider the example of *Petrofac*, a global oilfield services company. They revamped their contractor onboarding process in the Middle East to include not just task-specific training but hydration education, dietary advice, and access to shaded rest areas. The result? Fewer heat-related illnesses and a marked improvement in near-miss reporting, because contractors began to trust the system.

Or *Laing O'Rourke*, one of the UK's leading construction firms. They implemented a workforce well-being strategy that extended to all on site personnel, including agency workers and subcontractors. Initiatives included site-based mental health first aiders and anonymous feedback channels. The outcomes were clear: higher worker engagement scores, lower turnover, and, crucially, an uptick in proactive safety observations.

Or take *Saudi Aramco's* contractor management model. In their Safety Management System Manual (SMS), Aramco insists on a Contractor Health and Welfare Plan as a contractual requirement. This plan includes cooling rest shelters, health screenings, potable water access, and grievance mechanisms. Far from being a formality, this standard has helped the company manage some of the world's most complex megaprojects with a safety performance that rivals many Organisation for Economic Co-operation and Development (OECD) nations.

The takeaway is simple: **welfare is not a soft topic**. It's a hard metric of operational maturity. And it must be embedded—**not performative, not temporary, and not outsourced**.

Embedding means building these expectations into prequalification documents, into tender scoring, into onboarding, into audits, and into daily supervision. It means making contractor welfare a standing item in site leadership meetings. It means tracking complaints not just as issues to be closed, but as data to be learned from. It means ensuring every site leader knows that water temperature and bedding conditions are as much their concern as PPE compliance and permit controls.

When you embed safety and welfare into your **systems**, you create organizational resilience. Systems don't depend on goodwill or exceptional individuals. They create the *conditions* for predictable outcomes—especially under pressure. And that's what the best contractor management does. It doesn't simply issue instructions. It creates the environment where people can perform safely, repeatedly, and sustainably.

As we move forward in this book, remember these examples. They are not relics of the past. They are warnings from the present. The contractor may be external to your payroll, but their safety is **internal to your responsibility**. Their welfare is **your reputation**. Their performance is **your performance**.

Contractor safety is everyone's business—and never someone else's fault.

Why Does It Matter? 2

Why does it matter? For good commercial reasons as well as on ethical grounds. Strong Health, Safety, and Environment (HSE) contractor management improves productivity and reduces worker turnover as well as avoiding lost time.

Let's look at it in its simplest terms. You have your own standards and procedures. In all likelihood, you also have some other standards or regulations—or even laws—that you have to comply with. They cover everything your company does. But, the way things are now, everything your company does is not done by your company. An increasing proportion is done by contractors. If you leave it to the contractors to do things their way—perhaps by saying, "You're the experts, so we leave it to you"—how certain can you be that it will be done the way you would prefer it done? And, even more importantly, that it will be done as the law says it should be? If your contractor breaks the law while acting on your behalf under contract with you, who are the legal authorities going to go after? The first question they are likely to ask is: who has the money? If it's you, the client, it could be expensive not only in fines and the cost of bringing the work up to standard but also in lost work in the future. Specially if you find yourself banned from working in that same jurisdiction for a period of time. And that happens.

Your legal team may argue that responsibility can be shifted through robust contractual arrangements, with liabilities assigned clearly to contractors and financial risk passed on to insurers. To a limited extent, they would be right. Contracts can create legal distance, and insurance can help recover costs. But neither contracts nor policies will deflect the blow of public opinion. When something goes wrong, particularly in the realm of contractor welfare and safety, public scrutiny does not concern itself with fine legal print—it focuses on the brand.

A prime example of this was the global backlash faced by Nike during the 1990s and early 2000s over the treatment of workers in its subcontracted factories, particularly in Southeast Asia. Reports emerged that workers in factories producing Nike goods were subjected to poor wages, unsafe working conditions, and even physical abuse. Although these factories were owned

DOI: 10.1201/9781003754121-2

and managed by third-party contractors and not directly by Nike, that fact made little difference to the court of public opinion. Global protests were sparked, consumer boycotts were organized, and Nike's brand reputation suffered a major blow. Media coverage, including investigative pieces by CBS and the BBC, focused squarely on Nike, not on the subcontractors. The widespread perception was that Nike was complicit through inaction, failing to ensure its products were made under ethical and safe working conditions. This case served as a turning point, not just for Nike but for the broader corporate world, forcing many companies to rethink and reform their supply chain oversight and contractor welfare standards. It highlighted that reputational accountability does not respect the boundaries drawn by legal contracts.

The impact on Nike was immediate. Public outrage and consumer boycotts hit the brand in major markets like the US and UK. Though Nike had no legal control over subcontracted factory conditions, the public held the brand accountable. The Nike swoosh came to symbolize exploitation instead of performance. As a result, Nike had to overhaul its supply chain management by introducing factory audits, codes of conduct, and public transparency. This moment reshaped how multinational firms view contractor responsibility. The reputational fallout was a far greater cost than any legal liability, proving that, once public trust erodes, no contract or insurance policy can restore it.

This episode is an instructive case study for any client organization. It underscores the point that reputational risk is real, global, and often unforgiving. Stakeholders, from consumers to investors, expect contractor safety and welfare to be part of the client's responsibility, whatever the law says. And once public trust is lost, no clause in a contract and no payout from an insurer can buy it back.

We'll come back to this in future chapters, because it's a trend that everyone needs to be aware of. Increasingly (and look back at what we said about the government in Qatar refunding money workers had paid to get a job there, even though the money had not been paid to the government), what matters is not "Whose fault was it?" but "Who can afford to pay?"

Let's say you have contracted to build and operate a number of offshore wind farms. I don't want it to seem that I'm pointing fingers at anyone (and I'm not), so let's not be precise about the location. It's somewhere in Europe—could be the North Sea, could be the Baltic, the English Channel, or the Bay of Biscay.

You won't be doing all the construction work yourself. Building undersea foundations is specialist work. So is erecting turbines on top of them. Connecting the turbines to land so that you can transfer the power you generate will be best done by someone else again. I could go on; probably listing 50 or more completely separate operations from the time you start planning through the design, construction, and installation phases and then the operational phase that would best be done by specialist teams working for companies of whom it can be said: that's what they do. Each brings its own kind of risk which the contractor will have to manage, but making sure they manage it in accordance not only with your standard procedures but also with what the relevant statutes say falls, in the end, on you. And those legal considerations can be complex—it's possible to imagine the kind of project we are discussing here being subject to legislation in more than one jurisdiction. France and the UK, for example. Germany, the Netherlands, and Belgium. (Yes, all but one in that case are in the EU, but they have their own maritime regulations). And that is apart from international agreements and obligations summarized as the Law of the Sea which have a great deal to say about territorial waters, extraterritorial waters, and the rights of navigation.

Of those 50 or more possible operations that you might have subcontracted, I'm going to take one in particular. Operation of a guard vessel.

For those who haven't worked in this industry, cardinal buoys will be positioned along the edge of a safety zone around the site where an offshore wind farm is being erected. The size of the safety zone will have been agreed with the regulatory authorities for that stretch of sea. What a guard vessel does is to monitor what's happening inside that safety zone. Other vessels are going to come into the zone carrying construction workers and materials. The people on the guard vessel need to know at any time what vessels are expected; they need to monitor the vessels all the way to the site and ensure that they remain safe.

But that isn't guarding—that's escorting. The guard duties come into play when a vessel comes close to the safety zone or enters it unexpectedly. This is particularly likely to happen in stormy weather (and it sometimes seems that storms are the norm in the North Sea, the Baltic, and the Bay of Biscay) but it can also happen in areas with high levels of marine traffic such as the English Channel. And, if you happen to be doing this work right now in the South China Sea, there's a whole other set of possible intruders. Guarding by a guard vessel continues all the way through construction and commissioning and only ends when the wind farm has been handed over to the ultimate owner (in this example, your company). At that point, a whole new procedure for protecting the installations against intrusions (and protecting the intruders themselves) comes into play.

Someone has contracted with the company that owns and runs the guard vessel. That someone may not be your company—it's normal practice for the company in overall charge of the project to agree with the company constructing the turbines that the construction company will enter into a contract with the guard vessel's owners. But that still leaves the question of how you, as the ultimate customer, ensure that the safety procedures followed by the guard vessel's crew are aligned with your own safety procedures.

WHAT HAVE OFFSHORE WIND FARMS AND GUARD VESSELS TO TEACH THE REST OF US?

I imagine many people reading what I've just written and saying, "Well, that's very interesting, but I have nothing to do with offshore wind farms and I don't believe I will ever need to put a guard vessel on station. So what has any of this to do with me?"

The answer is that offshore wind farms are a high-risk area for both health and safety, and guard vessels are a security measure. And you do have an interest in health and safety risks and in security measures, because if you didn't you wouldn't be reading this book. So let's summarize the lessons we could have drawn just as easily from roadbuilding projects or office building projects or … Well, just about any kind of building project. And then we'll look at how they can be applied elsewhere.

THE PROJECT'S ENVIRONMENT AND WHAT WE MUST BE AWARE OF

The Regulatory Environment

The offshore wind farms regulatory environment is complex because of the number of sets of rules that apply. They include:

- International agreements and treaties known collectively as the Law of the Sea
- National laws
- Regional laws
- International agreements and commitments concerning health and safety entered into by all involved states and regions

Clearly, not all of these apply to all onshore projects—but many do, especially when a project is carried on in more than one legal jurisdiction.

The Physical Environment

Working offshore brings problems of its own—the question of tides; extremes of weather; the nature and state of the seafloor; underwater rocks and other hazards to mention but a few. But working onshore also has problems and they include the weather (and let's not forget hurricanes and tornadoes) and the nature and state of the subsoil. The latter will sometimes need extensive dewatering before work can start and while it continues; in others, it will mandate, for example, piling and/or other work to provide a sufficient level of stability. And then there is the question of the proneness or otherwise to earthquakes, and the need to protect structures against them. An additional set of problems can arise when the project is onshore but close to water. Is erosion likely to be a factor? If this is an earthquake zone, what are the possibilities of a tsunami? What about rights people have to walk, drive, or otherwise comport themselves on or across your site? How is that going to affect your ability to keep the place safe? And have you looked at the history of the site? In Shropshire, England, a building contractor got approval to build new homes in 2020 on what until the 1950s had been a Ministry of Defence base. One large field on which many of the new homes were to be built had been an artillery training area. There should have been a survey using, at the least, ground penetrating radar to identify hazards. No such survey took place. After construction had begun, it was repeatedly halted because unexploded shells were turning up as foundations were built. The repeated delays this caused made an enormous hole in the profits the construction company had expected, and also made people very wary of buying one of the new houses—how would you like to be turning over your new garden and hit an unexploded shell? The lesson: you need a risk analysis and it must cover every possible risk and go into sufficient depth.

The Working Environment

Anyone who has been involved in more than only one or two construction projects knows that some go swimmingly from the start and some cause difficulties from the very beginning. A lot of this has to do with things we've already talked about: the physical and mental health of the people working on the project. And as we've said, that's something you can do something about. But some is to do with the care you took when you were setting the project up. Which leads us to:

THE PROJECT DOCUMENTATION—AND HOW YOU CREATE IT

Often referred to as the project manual, you need a document that covers every conceivable aspect of the work to be done. In the case of the offshore wind farm mentioned above, there would be a section for undersea exploration, one for building the foundations, one for keeping the area safe, one for communications, one for building the structures, one for the layout and construction of the control room—and so on. For an offshore wind farm, there might be as many as 50 sections. Each section should include as an absolute minimum:

- Who is the client for that part of the project?
- What outcomes is the client entitled to expect?
- Who is responsible for meeting those expectations?
- What risks have been identified, how will each of them be dealt with and who is responsible in each case?
- What steps can be taken to preserve the health and safety of all members of the workforce?
- What steps can be taken to preserve the well-being of all members of the workforce?

In addition, each section should have its own Gantt or similar charts together with drawings of the proposed work and the layout of each portion.

This can only be done effectively if committees are formed for each part of the project and the committees comprise people from both sides—the clients awaiting delivery, and the contractors responsible for delivering. They must contribute based on their own previous experience and on risks and considerations related specifically to this project.

There is often a temptation to get started without having completed the work needed to produce this document. It should be resisted. Any time you start a project of significant size without first completing the complete set of documentation for the project, you will end up with costs higher than they needed to be.

And, as the project proceeds, further meetings will be needed to compare how the work is developing to how it was expected to develop, and to amend the documentation as needed. Project documents should have a revision number and date and someone should have responsibility for ensuring that everyone is working to the most recent, most current version.

It's important to note that some of the clearest, most practical examples of contractor safety management processes have come from the oil, gas, and petrochemical sectors. These are industries where the consequences of failure can be catastrophic—not only in terms of immediate fatalities or environmental damage, but also in terms of reputational and legal impact that can last decades. The US Chemical Safety and Hazard Investigation Board (CSB) has investigated numerous incidents where poor contractor oversight has led to disastrous results.

One well-documented case is the 2005 BP Texas City refinery explosion. Fifteen workers were killed and over 170 injured when a hydrocarbon vapor cloud ignited during the startup of a distillation tower and raffinate splitter. Many of the victims were contractors working in trailers positioned too close to the process units. The CSB investigation highlighted that BP's contractor management systems were poorly enforced. Safety training was inconsistent, there was little or no communication about ongoing high-risk activities, and inadequate hazard analysis also contributed to the disaster. BP faced widespread criticism not only for the technical failings but for a broader cultural failure to integrate contractors into its safety systems.

Another example is the 2010 Tesoro refinery explosion in Anacortes, Washington. Seven workers were killed when a heat exchanger ruptured due to high-temperature hydrogen attack. The CSB report noted that Tesoro relied heavily on contractors for mechanical integrity assessments, but lacked rigorous processes to evaluate the quality of their work or ensure alignment with internal safety standards. As with Texas City, it wasn't just the contractor's failure—it was the client's failure to adequately oversee and integrate the contractor's activities.

Both cases underscore the need for strong contractor engagement frameworks. In response to incidents like these, the International Association of Oil & Gas Producers (IOGP) has issued comprehensive guidelines on contractor safety management. These publicly available documents offer practical steps for prequalification, safety onboarding, performance monitoring, and ongoing engagement.

Organizations outside of oil and gas—particularly those in construction, infrastructure, and renewables—can benefit from studying and adapting IOGP's guidance. Whether your project is an offshore wind farm or an urban high-rise, the principles remain the same: clear ownership of safety, continuous collaboration, and transparent expectations between clients and contractors. Using frameworks like those recommended by the IOGP not only enhances safety outcomes but also helps organizations show due diligence and mitigate reputational risks.

In short, there's no excuse for not having a solid contractor safety system in place. The resources exist. The examples—both of failure and of improvement—are clear. And the consequences of getting it wrong are more public and more permanent than ever before.

PROCESS SAFETY MANAGEMENT

Another area where contractor management plays a critical role is in regulating high-risk operations, particularly within the process industries such as oil refining, petrochemicals, chemical manufacturing, and gas processing. These industries share the potential for catastrophic accidents because they all involve handling hazardous materials under high pressures and temperatures. Because of the high stakes, the consequences of a misstep in these environments can be far-reaching: injuries or fatalities, environmental disasters, loss of property, and severe reputational damage.

To mitigate these risks, many jurisdictions have developed and enforced a framework known as Process Safety Management (PSM). PSM is a disciplined, structured approach to managing the integrity of operating systems and processes that handle hazardous substances. The framework is not a checklist—it is a cultural and technical system that aims to prevent unplanned releases of substances that could cause serious harm to workers, the public, or the environment.

In the United States, the Occupational Safety and Health Administration (OSHA) regulates PSM through its standard 29 CFR 1910.119. This standard outlines 14 key elements that must be addressed when dealing with processes involving highly hazardous chemicals (HHCs). Among these core elements, contractor management is specifically included—not as a side note, but as a central tenet.

OSHA mandates that host employers must obtain and evaluate the safety performance and programs of contractors, ensure that contractors are informed about known hazards, and maintain active oversight of any contractor operations that could impact process safety. This requirement reflects a broader operational truth: in high-risk industries, contractors are often engaged in some of the most dangerous tasks—shutdowns, equipment repair, inspection, and turnarounds. If these contractors are not fully integrated into the client's process safety framework, the potential for failure escalates sharply.

A case in point is the 1989 Phillips 66 explosion in Pasadena, Texas. During a routine maintenance operation at a polyethylene plant, a release of highly flammable process gases led to a series of explosions, killing 23 workers and injuring more than 130. Many of the deceased were contractor employees who were either not properly briefed on hazards or were left vulnerable due to inadequate emergency preparedness. Investigations by OSHA and the CSB uncovered serious lapses in communication, contractor training, and process hazard analysis. Inadequate lockout/tagout procedures and rushed work schedules further compounded the danger. This incident remains one of the most haunting examples of how failure to integrate contractors into PSM systems can have fatal consequences.

The 2009 Silver Eagle Refinery explosion in Utah also underscores these issues. A hydrogen pipeline rupture caused a massive explosion that injured four people and caused extensive damage to nearby homes. Investigators found that insufficient mechanical integrity protocols and poor hazard analysis practices, especially in relation to contractor activity, were contributing factors. Contractors engaged in maintenance work were neither properly trained nor fully integrated into the refinery's PSM systems.

Such incidents highlight the need for a contractor management program that is robust, formalized, and integrated into the broader process safety framework. Simply awarding a contract based on lowest price or basic qualifications is not enough. In process industries, the contractor is not a visitor—they are often responsible for executing the riskiest aspects of the job.

Leading organizations now incorporate contractor PSM alignment as a requirement in their procurement and onboarding processes. Shell, for example, requires that all contractors engaged in high-risk operations undergo specific PSM training that mirrors the internal protocols followed by Shell employees. This ensures a common understanding of the risks, controls, and behavioral expectations. Dow Chemical's "Contractor Safety Management Guideline" incorporates a five-stage contractor safety lifecycle—from prequalification and selection through to post-job evaluation. Contractors are evaluated not just on their safety record but on their safety culture, supervisory competence, and willingness to participate in joint safety planning.

In the European Union, the Seveso III Directive mandates that companies managing large quantities of dangerous substances must have detailed safety reports and emergency plans. The directive places specific responsibilities on operators regarding third-party and contractor involvement. In 2015, the Buncefield oil depot explosion in the UK prompted further strengthening of these provisions. Investigations found that poor communication between contractors and operators during a tank filling operation was a key cause of the incident.

Contractor management also plays a vital role during turnaround and shutdown operations—periods that are statistically the most dangerous for process facilities. These operations often involve hundreds of temporary contractor personnel, working around the clock under tight deadlines. The Centre for Chemical Process Safety (CCPS) recommends implementing pre-turnaround contractor safety audits, mandatory onboarding sessions, and cross-functional coordination meetings involving operations, safety, and contractor supervisors. In 2010, during a turnaround at a large Gulf Coast refinery, Chevron implemented a shared permit-to-work system that included contractor leads and enabled real-time visibility of hazardous work across the facility. The initiative significantly reduced simultaneous operations risks and contributed to the event being free of recordable incidents.

Another key focus area is the auditing and performance monitoring of contractors. Effective PSM implementation includes tracking, leading, and lagging indicators specific to contractor safety. These may include stop-work interventions, participation in safety meetings, and proactive hazard reporting. Fluor Corporation, one of the largest global engineering and construction firms, uses a proprietary system called Zero Incident Performance (ZIP) to integrate contractor and employee safety data. The system enables predictive analytics to identify risk trends in contractor behavior before incidents occur.

The importance of contractor management in process safety is recognized globally. The International Association of Oil & Gas Producers (IOGP) publishes guidelines specifically targeting contractor safety in high-risk industries, emphasizing shared accountability, integrated planning, and continuous improvement.

The lesson from years of both success and failure is unequivocal: contractor management cannot be an afterthought in process safety. The risks are too great, the consequences too severe, and the public scrutiny too intense. Organizations that treat contractor integration as a core pillar of PSM are the ones that succeed—not only in preventing disasters but in building a sustainable, high-trust safety culture.

3 Global Best Practices in HSE Contractor Management

We talk about global best practices, and we'll continue to talk about them, but it has to be recognized that culture and the regional economy have a profound effect on attitudes to Health, Safety, and Environment (HSE) and on the level of knowledge. As we've already discussed, this cannot be placed in the contractor's hands, attractive as that option may appear. Safety, in addition to sustainable operations leadership and focus, are the client's responsibility.

At first sight, it may look as though globalization would lead to a fall in standards everywhere as companies had to cut costs in order to fight competition from cheaper competitors. In practice, that has turned out not to be true and the reasons have a great deal to do with another aspect of globalization: what you do in one place is now a lot more visible everywhere else.

However, where cost-cutting drives the selection of contractors, and oversight is minimal, the results can be disastrous. Consider the 2010 Deepwater Horizon oil rig explosion in the Gulf of Mexico, operated by Transocean under contract for BP. Eleven workers lost their lives, and millions of barrels of oil spilled into the ocean. Investigations found that a series of cost-saving decisions and a failure to integrate safety practices between BP and its contractors played a central role in the catastrophe. Despite being a leading multinational, BP faced immense reputational and financial fallout, including over $60 billion in fines and settlements.

Another example is the 2007 collapse of the I-35W bridge in Minneapolis, USA. While the root cause was ultimately traced to design flaws, failing to properly oversee contracted maintenance and inspection contributed significantly to the tragedy. The Minnesota Department of Transportation had outsourced regular inspection and assessment activities to private firms.

DOI: 10.1201/9781003754121-3

Following the incident, it was revealed that outdated calculations and poor documentation had gone unchecked. Thirteen people were killed, and over 100 injured. The collapse became a symbol of how trying to save on upfront costs can lead to vastly greater long term liabilities.

Both examples underscore that saving money in the short term by engaging lower-cost contractors without the proper management frameworks can cause human loss, regulatory penalties, reputational damage and, ultimately, far higher financial costs than if best practice had been followed from the outset.

It's now over four decades since what is known almost universally as the Bhopal disaster. In December 1984, a pesticide plant operated by Union Carbide India Ltd in which the American company Union Carbide held a majority stake leaked methyl isocyanate gas into the surrounding community in Bhopal, India. The immediate result was the death of thousands, with conservative estimates citing over 4,000 fatalities and some reports suggesting a toll exceeding 15,000. More than half a million people were exposed to the gas, resulting in long-term health impacts ranging from respiratory illnesses to neurological and psychological disorders. The tragedy remains one of the worst industrial disasters in history.

The aftermath of Bhopal highlighted multiple failures in industrial safety, including poor equipment maintenance, inadequate emergency response systems, and a lack of transparency. More importantly for this discussion, it exposed a systemic failure in contractor and subsidiary oversight. Union Carbide maintained that plant operations were the responsibility of its Indian subsidiary, but this argument did little to shield the company from global outrage and reputational damage. The site remained contaminated for decades, and only in January 2025, after years of pressure from activists and legal campaigns, did Indian authorities begin the cleanup of over 300 metric tonnes of hazardous waste.

We are not going into detail about the protracted negotiations and series of accusations flying about in both directions after the leak. Union Carbide is now owned by Dow Chemical and to this day maintains that it did nothing wrong and that the disaster was caused by employee sabotage. Lawyers representing the victims say that maintenance cuts and other financial shortcomings meant that the equipment was not up to the job. There was a cash settlement that Union Carbide claims was more than sufficient and lawyers representing the victims say was woefully inadequate. That is not what is under discussion here—we don't know enough about the technicalities of what happened to say that one side is right, the other side is right, both sides are right, or neither side is right. What we do know is that the aftermath of a disaster like this would be very different today. And it was to some extent the Bhopal disaster that kick-started the changes.

Everything that happens now is more highly visible than it used to be. Governments that one might have considered invulnerable have been toppled by movements that drew worldwide attention to abuses they either promoted or tolerated. It is not possible to ignore the agency of human rights organizations in these changes.

No-one, and probably not even Nike, would really deny that Nike in the 1990s did not treat its Indonesian workforce as well as it might have done. It operated what amounted to a sweatshop operation. To some extent, it was the Bhopal disaster that motivated human rights organizations to be vociferous in their condemnation of abusive work practices. What made the difference was that they were now able to persuade governments to take the matter seriously.

Two governments were involved here: those of America and Indonesia. Both were motivated by a wish to avoid negative publicity. The American government in particular certainly wanted to make sure that the insults that had been thrown in America's direction after Bhopal were not repeated. That is the power of reputational damage, and if a reader of this book takes away nothing else, they should take away the importance of reputational damage.

The United States government said that, unless the Indonesian government did something to combat human rights abuses, Indonesian exports would lose the tariff privileges they enjoyed. The Indonesian government promptly raised the minimum wage, in effect increasing it by 400% in a very short period of time. This was the perfect moment for the classic argument about the weakness of globalization to prove itself: according to globalization's critics, the number of people employed at the bottom end of the pay scale in Indonesia would now fall as companies moved production to countries where labor costs were lower.

That is not what happened. While detailed statistics are limited, reports from the period suggest that many companies, under pressure from public scrutiny and governmental oversight, continued operations in Indonesia, leading to the maintenance—and in some cases growth—of low-wage manufacturing jobs. For example, a 2001 Global Alliance for Workers and Communities report covering Nike-contracted factories found wage increases and modest improvements in worker conditions. Although causation is difficult to prove definitively, fears of mass job loss due to improved labor regulations had been overstated. What this shows is that raising standards and managing contractors responsibly does not necessarily lead to a loss of opportunity—it can, in fact, coincide with a more stable, transparent, and sustainable employment environment.

Something else that went on at that time was that human rights organizations began to highlight the concern by publishing numerous articles about abuses in employment in developing countries by companies from more highly developed nations. Activists realized they could shame companies into doing better—not because they were embarrassed to be shown up in print, but because people took note of what they read and boycotted their products.

Why are we telling you this in such detail? Because that's exactly what can happen to you if you don't take notice. There was a time when big, powerful companies could simply ignore negative publicity. Ask Union Carbide. That time has gone. Ask Nike. Because Nike's Korean managers in Indonesia had negotiated with the Indonesian government a special exemption from the minimum wage legislation. And the negative publicity that brought down on the company's head was causing so much damage to its sales that the company had first to ignore its exemption and raise wages across the board and second to spend an enormous amount of money and management time on presenting a more positive picture to the world.

It would be nice to think that every company using contractors would employ the very best practices on nothing but ethical grounds. The reality is that companies understand that failure to employ best practice can be extremely expensive.

The fact that standards in the country where you're about to sign a contract may be lower than you're used to is no excuse for neglecting best practices. Using that as justification to cut corners or treat local people and environments as expendable is a fast-track to long-term reputational damage—as history has shown in cases like Bhopal. Reputations built over decades can be undone in a moment, and the consequences may linger far beyond the project's duration. But do not fall into the opposite trap either by assuming that the people in a less economically developed country lack the training, skill, or professionalism required. In many cases, the local workforce is highly capable. What they need is leadership that respects and reinforces standards instead of assuming incompetence or imposing second-rate conditions.

THE MIDDLE EAST

In Chapter 1, we talked about migrant workforces in the Middle East, and whether the death rate among them is fairly reported. (It clearly is not—but where does that get us? Everything so far in this chapter has been about the importance of publicity and the reputational damage it causes. Presenting the opposite case when a negative claim has been made about you takes time, and people often won't give you that. "6500 non-native construction workers have died in Qatar since Qatar was awarded the FIFA World Cup" is a catchy headline. For many people, it's all they'll read. You could respond by saying, "No, no, only 30,000 migrant workers are employed on the World Cup while the total number of migrant construction workers in Qatar is two million, and that's where the 6500 came from." Yes, you could say that. And it would be true. But if no-one is listening, what good does it do?)

The claim of 6,500 deaths, widely circulated by major outlets such as The Guardian, was based on data compiled from five South Asian countries over a 10-year period. However, the report itself admitted that the deaths were not necessarily linked to World Cup projects, nor were they all construction-related. The figure included all causes of death—illness, natural causes, accidents—across all sectors, from domestic work to retail to infrastructure.

When official figures were finally published by Qatar's Supreme Committee for Delivery & Legacy, it was confirmed that **there were three work-related fatalities and 37 non-work-related deaths among workers directly involved in World Cup projects**. These numbers, while tragic in their own right, are significantly different from the figures being discussed in global media. Moreover, these fatalities occurred over more than a decade across a massive infrastructure push, including stadiums, metro lines, highways, and hospitality zones.

This discrepancy wasn't just about numbers. It became a lesson in narrative control, perception, and the Western media's selective framing. The clarification by Qatari authorities and FIFA's own reporting received limited coverage. Corrections—when they came at all—were buried deep within articles.

For the international public, the damage was done. The headline stuck. And in today's media landscape, **headlines often win over facts**.

The takeaway for major projects and host nations is clear: it's not just what happens, but how quickly and transparently you communicate it—and how loudly. Silence or late response will leave a vacuum that others will fill. In the case of Qatar, an entire decade of worker welfare reform—including heat stress legislation, wage protection systems, and accommodation standards—was overshadowed by one number that, although technically true in a broader demographic sense, was **contextually misleading**.

For contractor management professionals, the Qatar World Cup controversy is a powerful reminder that **project success depends not only on safety performance but on storytelling, data literacy, and the ability to proactively manage public perception.**

Since then, Qatar has implemented a range of reforms in recent years aimed at improving conditions for migrant workers, including abolishing the Kafala (sponsorship-based labor) system, introducing a non-discriminatory minimum wage in 2021, and establishing labor dispute committees and heat stress protection measures. Similar efforts have been made across the Gulf

Cooperation Council (GCC) states. The UAE has introduced reforms promoting wage protection and mandatory rest periods, while Saudi Arabia launched its Labor Reform Initiative to improve job mobility and transparency in employment contracts. Kuwait and Oman have also updated their labor laws to better protect foreign workers' rights. These reforms, many of which are informed by international labor standards and have been praised by organizations such as the International Labor Organization (ILO), mark significant progress. In some cases, the regulatory protections for migrant construction workers in these states now meet or exceed those found in parts of the Western World—particularly in terms of legal clarity and enforcement mechanisms. This is why they are increasingly cited as examples of emerging global best practice.

We are not going to list every single regulation you have to follow to work in one of these countries, because that is not the point: you will, of necessity, be checking what the regulations are in any country you intend to operate in, and you will be looking for ways to ensure that they are complied with. What we are going to do is to list a few to give a feeling for what "best practice" means.

Recruitment Fees

If a worker leaves their family and all they know in Tamil Nadu, Bangladesh, Nepal, or Sri Lanka to work for minimum wage in one of the oil-rich Middle Eastern countries, you can be pretty sure they aren't doing it because of an innate love of travel. They are doing it because they have to—the job opportunities available to them at home, if any even exist, are scarcely enough (or, sometimes, not enough) to keep them alive. They are prepared to pay a recruitment fee to get the job. Almost without exception, those fees are collected by local agents in the place where they live and it is those agents who set the fee. It isn't something that is passed on to the company they are going to be working for. In the past, companies ignored the knowledge that some people who traveled great distances to work for them had been fleeced by agents charging large sums to let the person join the work program. The companies felt this was not their business. Increasingly, they are not allowed to ignore it and some construction companies are now reimbursing migrant workers with at least part of the money they paid to get there.

This issue is not only a humanitarian concern but one with serious legal, reputational, and operational implications for contractor management. The United Nations' International Labour Organization (ILO) categorically states that recruitment fees should be borne by the employer, not the worker. The ILO Fair Recruitment Initiative, launched in 2014, defines employer-pays models as the ethical standard. Yet widespread violations persist, particularly in the construction, hospitality, and domestic work sectors in the Gulf and Southeast Asia.

One illustrative case is that is of **Turner Construction's efforts on the New York University (NYU) Abu Dhabi campus project**, which drew scrutiny after allegations surfaced that migrant workers had paid large fees to secure employment. An independent review conducted by Nardello & Co., commissioned by NYU, found that some workers paid fees equivalent to several months' wages. In response, NYU worked with contractors to identify the recruitment agents involved, pressurized them to cease unethical practices, and committed to reimbursement for affected workers. The university also revised its labor compliance protocols to avoid future lapses, setting a precedent for educational institutions acting as clients.

In **Malaysia's electronics sector**, a 2020 report by Verité, a US-based non-profit, highlighted that nearly 80% of surveyed migrant workers had paid recruitment fees—often resulting in debt bondage. Companies such as Intel and HP responded by adopting Employer Pays Principles and demanded that their Malaysian suppliers refund workers. This push has since influenced supply chains across the region, linking ethical recruitment to global procurement standards.

These examples illustrate a fundamental truth in contractor management today: your responsibility doesn't start when the worker walks through the gate—it begins when they consider applying.

For clients, this has several implications:

1. **Due Diligence at Source**: Contractor prequalification should include a clear review of their recruitment channels. Are third-party agents used? Are recruitment practices audited in the country of origin?
2. **Contract Clauses**: Insert clauses that prohibit worker-paid recruitment fees and require proof of zero-fee hiring practices. Include financial penalties for violations and remediation requirements.
3. **Worker Interviews and Audits**: Conduct worker interviews during site inspections and audits to identify any recruitment-related debt or coercion.
4. **Reimbursement Programs**: Establish procedures for reimbursing workers if unethical recruitment practices are uncovered. Clients should consider a joint responsibility model, where the contractor initiates payment and the client matches it or supports through project funds.

5. **Awareness Campaigns**: Support programs that educate migrant workers on their rights, ideally in their native language, and create anonymous channels for raising concerns.

But beyond process and policy, there is a moral dimension here that cannot be overlooked.

These workers are often the sole breadwinners for extended families. A $1,500 recruitment fee might represent multiple years of savings or drive them into high-interest loans. By the time they reach the job site, they are not merely poor—they are indebted. And debt creates pressure. It leads workers to cut corners, to accept unsafe conditions, and to remain silent when risks emerge—out of fear of being sent home or losing their ability to repay loans.

This makes recruitment fees not just a welfare issue, but a direct safety risk.

For example, in a **construction project in Bahrain** managed by a European engineering firm, site safety auditors found that workers were skipping Personal Protection Equipment (PPE) usage and working overtime without breaks. Upon investigation, it emerged that the majority had paid recruitment fees ranging between $1,000–$2,500. In interviews, workers admitted they were trying to earn back the debt quickly, hence the long hours and reluctance to raise safety concerns. Once the firm introduced a reimbursement program and financial counselling, incident rates dropped, and compliance with PPE protocols rose significantly within three months.

Contractor management teams must treat this issue with the seriousness it deserves. It is no longer sufficient to say, "We didn't know." Clients and main contractors must know—and must act. This includes:

- Partnering with NGOs and ILO taskforces to map recruitment corridors.
- Engaging third-party compliance monitors to conduct recruitment audits.
- Using technology such as blockchain-based employment contracts to verify the financial transaction history of hired workers.

Some governments are taking steps, too. **The Philippines government**, for instance, has strengthened its licensing requirements for recruitment agencies and established monitoring frameworks to ensure fair recruitment. Through its partnership with the International Organization for Migration (IOM), the country promotes the ethical recruitment of overseas Filipino workers (OFWs) and penalizes agencies that impose illegal fees.

But these mechanisms only work if clients and contractor teams use them. Ignorance is not a defense.

Ultimately, the elimination of recruitment fees is more than a compliance box to be ticked—it's a litmus test for leadership. The companies and project owners who actively tackle this issue are not just reducing risk. They are transforming lives.

And if that's not part of your HSE vision, maybe you need to broaden it.

ISO 45001

This is the International Health and Safety Standard. It formally replaced OHSAS 18001 as the international standard for occupational health and safety management, introducing a more proactive, risk-based approach that required organizations to transition by March 2021 to maintain certification.

The procedure to be certified as ISO 45001 compliant is rigorous and confirms that a company meets high standards of occupational health and safety management. One way to ensure that your contractors operate in line with your expectations is to require them to be ISO 45001 certified. However, this should not be taken at face value. Always verify the certification—and not just its presence, but its validity. Ensure it has been issued by a reputable and accredited certification body, ideally one recognized by international accreditation forums such as the International Accreditation Forum (IAF) or the relevant national accreditation authority. A certification issued by an unknown or unaccredited entity may carry little to no credibility. It is vital to check not only that the certification exists, but that it is current, authentic, and issued by a trustworthy provider. You don't want to discover after a serious incident that the ISO 45001 certification you relied on was invalid or misrepresented.

The essential point of ISO 45001 is that companies complying with it have converted risks to controlled risks. A vital point of risk management that is forgotten far too often is that it is, indeed, management of risk. A surprising number of people seem to believe that, if you have managed risk, you have avoided it and that is not true. Risk is risk; its essence is that the event of which a risk exists can happen, however well managed it is.

So how can ISO 45001 certification truly be leveraged in contractor management? And how can clients ensure they're using it not just as a compliance checkbox, but as a real-world tool for proactive safety assurance?

First, ISO 45001 is based on the Plan-Do-Check-Act (PDCA) cycle—a systematic process for continuous improvement. When evaluating a contractor, clients should require more than a copy of the certificate. They should request a description of the PDCA process as it applies to that contractor's operations. Is the contractor identifying hazards systematically?

Are they reviewing risk registers at routine intervals? Do they engage workers in risk evaluations? These are not just audit questions—they're cultural indicators.

Balfour Beatty, one of the largest construction and infrastructure firms in the UK, applies ISO 45001 not only internally but across its supply chain. Contractors must undergo a safety competency assessment that includes ISO 45001 certification—but they are also expected to align with Balfour's internal safety philosophy, known as "Zero Harm." This includes demonstrating active worker engagement, a near-miss reporting culture, and leading indicator tracking.

What's interesting is how Balfour Beatty uses ISO 45001 **not as a shield but as a bridge**—a way to open conversations about safety governance. In one recent rail infrastructure project in Wales, a sub-tier contractor had certification but was found to be weak in incident trend analysis. Rather than drop them, Balfour provided guidance workshops to raise maturity. Certification was seen as a starting point—not an end point.

Malaysia's national oil and gas company, Petronas, requires ISO 45001 certification for all major contractors. But it doesn't stop there. The company uses its Contractor Health, Safety and Environment Management System (CHSEMS) to monitor contractor performance in real-time. Certification is verified at onboarding, but then tested through ongoing audits and performance KPIs such as LTI rates, PTW compliance, and safety meeting attendance.

In one offshore maintenance contract in 2021, Petronas noticed a divergence between certified processes and actual site behavior. Using ISO 45001 as a benchmark, the contractor's internal audit scores were compared against Petronas' external findings. This triggered a targeted improvement plan jointly developed by both parties. The result: a 43% drop in recordable incidents over the next 12 months.

The takeaway here is that ISO 45001 can serve as both a filter and a framework. It helps you screen who you let in, and guides how you work with them once they're in.

GVK Group's (An Indian Multinational Infrastructure Conglomerate) approach to safety during the Terminal 2 expansion of Chhatrapati Shivaji Maharaj International Airport (CSMIA) in Mumbai is a classic example of ISO 45001 implementation at the heart of contractor control. ISO certification was required not just from primary contractors, but also subcontractors in high-risk trades. However, what differentiated GVK's program was the **validation audits**.

The client's internal HSE team ran unannounced audits with third-party support, comparing actual work practices against the documented safety management systems. Subcontractors with valid ISO 45001 certification were flagged when field conditions—like unguarded scaffolds or incomplete toolbox talks—failed to match documentation.

As a corrective measure, GVK introduced a training module for certified contractors titled, "ISO in Practice," which walked teams through what compliance looks like on-site—not just in paperwork. Safety performance improved, and the initiative was later used as a model by other Indian airport projects.

There's a temptation to treat certification as a seal of perfection. But true due diligence demands more. In many high-risk industries, the gap between "audit-time compliance" and "real-time operations" is significant. Contractors may excel at preparing files for audits but falter on the shop floor.

To mitigate this, clients should:

- Verify the issuing body's accreditation with IAF or the national accreditation authority.
- Review recent surveillance or renewal audits.
- Interview HSE leadership about ISO integration into operations.
- Crosscheck third-party audit results with leading indicators like training completion, hazard reports, and toolbox engagement.

Another important dynamic is cultural. A contractor that pursues ISO 45001 certification proactively—not just when required—may already have a healthier safety culture. Their motivations matter. Are they doing it for the certificate or for the system?

A 2022 report by the **ILO** found that contractors who embedded ISO 45001 into their internal incentive programs (e.g., linking safety KPIs to bonuses) saw more sustainable safety performance than those who viewed certification as a contractual hoop to jump through.

In 2019, a subcontractor on a Middle East logistics project presented an ISO 45001 certificate that was later discovered to be issued by an unaccredited overseas body. It wasn't even legally valid. The incident only came to light following a fatality investigation. The parent company faced severe reputational backlash and incurred project delays due to the re-audit process.

This Is a Cautionary Tale for All Clients: Check, Verify, Validate. Trust—but Verify

ISO 45001 is more than a certificate—it's a culture framework. When applied properly, it helps clients filter, monitor, and elevate contractor safety performance. But used lazily or uncritically, it becomes a false sense of security.

Smart clients use ISO 45001 not just to ensure compliance—but to drive transformation.

The best contractor partnerships aren't based on certificates. They're based on mutual commitment to safety excellence, operational transparency, and shared accountability for risk.

Sustainability

Sustainability has been a buzzword for some time now, but for most of its currency it has dealt with resources: how to make the most of what we have, and how to make what we have last as long as possible. Things have been changing, because that's what things do, the fundamental nature of the world: it changes. Environmental sustainability has now been joined by social sustainability, and social sustainability means paying attention and not just lip service to workers' health and welfare as well as to workers' safety. The Middle East has seen the introduction and rise in importance of the Workers' Charter and the Workers' Welfare Standards.

But this is not confined to the Middle East. Let's take a specific example that happened in eastern Europe when a Western European company contracted with a Chinese construction company to build a cement plant in Albania. The Chinese company had responsibility for bringing in 700 workers and repatriating them when the work was over. Recruiting the workers was the responsibility of the Chinese, but the contract did not leave everything to them. The client developed a code of conduct that set out all the requirements the Chinese contractor was required to meet. These included, as a matter of course, compliance with ILO conventions and also Albanian and EU labor law—Albania was not at the time and still is not a member of the EU, but the client is subject to EU regulations and the fact that the plant was being built outside the EU does not change that. The code of conduct also included plans for supervising and monitoring the construction work. This covered everything imaginable connected with a 21st century construction contract; the points of particular interest as far as this book is concerned were:

- that safety training courses were in operation and all employees received them;
- that best practices for ensuring safety on-site were in force;
- that accommodation met standards laid down in the code;
- that food provided for workers was adequate and met at least the minimum requirements set down in the code;
- that workers had access to healthcare;

- that all accidents, from the most minor to major accidents, were logged and investigated, that first-aid was provided by designated and qualified first-aiders, and that subsequent care was provided as needed;
- that supplies of drinking water were available at convenient points on-site.

Once a company has signed an agreement like that, it is important to ensure that the terms are observed. It isn't necessarily the case that the Chinese contractor would have had all of those points in their own standard operating procedures. It isn't necessarily the case that they would all be standard Albanian practices. What that means is that there is a risk that some people involved in the contract may regard some of these provisions as window dressing—they look nice, and you can talk about them, but you don't actually have to follow them.

It's up to the client to ensure that ideas like that don't hold sway. And that means that it's up to the client to carry out regular checks. There are supposed to be supplies of drinking water at convenient points. Are they there? Every accident, however minor, is supposed to be resolved and investigated and first-aid is supposed to be provided by a designated and qualified first-aider. So look at the log. Is it being kept up to date? Pick out a logged accident. Where is the evidence that its causes have been investigated? Ask to see the designated first-aiders. Then ask to see their qualifications. And do some simple calculations—how many people work on this site and how many first-aiders are present at any given time? What is the ratio of designated, qualified first-aiders to workers on-site? Does it seem adequate to you?

Then go and look at the accommodation. Does it provide basic comfort? Are there any bare electrical wires and connections? Are there places where workers who are off duty can make themselves a drink and a meal and talk in comfort to other workers?

It's a fairly basic principle: a contract that isn't regularly checked is a contract that is likely to be broken. And if that breach ends up costing someone money, the someone is likely to be the client, because it's usually the client who has the money.

Welfare

Many of the allegations of abuse that have been leveled against construction companies in the Middle East have had to do with welfare. Accommodation standards were too low; people were expected to work long hours in high temperatures without adequate access to water; both physical and mental healthcare were neglected—and so on. These allegations have been met by putting in place legislation and other regulatory standards on welfare that would shame many Western countries.

WHAT THIS MEANS IN PRACTICE

So far, most of this section could, if you wanted, be treated as apple pie issues. Something that's nice to have but even nicer to talk about. If you want to follow global best practice, that won't do. Here are some of the things that the good contractor is putting into practice. They're not applicable only to the Middle East, but they are especially important in areas where temperatures are either very high or very low if you're planning a project in Canada's Northwest Territories, the Yukon or Nunavut, the meteorological and environmental challenges are likely to be every bit as great as in Saudi Arabia, though at the opposite end of the temperature scale (and don't forget the threat from bears).

Healthcare

If either you or one of your contractors employ enough people, it's a statistical certainty that some of them will fall ill. That's not a failure; it's a fact of life. But what is within your control—and absolutely within your responsibility—is how easy it is for them to access quality medical care when that happens. Can they be diagnosed accurately and promptly? Can they receive appropriate treatment, obtain prescribed medication, and have somewhere to rest and recover that doesn't make their condition worse?

These aren't abstract concerns. They are core operational realities. Too many organizations still treat medical care for workers—especially contractors—as a peripheral service or a mere regulatory checkbox. The mindset must shift. In any environment where physical labor is performed, where exposure to risk is present, and where living arrangements are often controlled by the employer, healthcare is not a perk. It's a pillar.

Let's start with the basics: are there enough trained first-aiders? Do they know how to handle the typical injuries associated with your work environment? Is there a first-aid kit in every zone on the site, and is it restocked regularly? Is there an emergency response plan, and has it been rehearsed? Have all parties on site—client and contractor—been briefed on who does what when something goes wrong?

Now let's go further. If you're operating in a remote location or in a high-risk industry, what's the escalation pathway beyond first aid? Is there a paramedic on site? Is there an occupational health nurse during critical working hours? Do you have a memorandum of understanding with a local hospital or clinic? If the answer to any of these is no, the next question is: why not?

We're not talking about hypotheticals here. The consequences of failing to provide adequate healthcare infrastructure are well documented. A 2022 study published in the *International Journal of Environmental Research and Public Health* found that construction workers operating in isolated environments—particularly in developing nations—suffered higher rates of untreated injury complications, prolonged absenteeism, and even mortality due to delayed or inadequate access to healthcare. One of the key findings? Contractors often lacked basic access to medical services while clients operated under the assumption that "someone else was taking care of it."

That is not good enough.

In the Gulf, the story is often similar but more insidious. Workers fall ill—not from catastrophic accidents but from chronic heat stress, dehydration, untreated infections, and long-term musculoskeletal problems. In 2021, *Human Rights Watch* documented how many migrant workers in the region lacked access to adequate healthcare despite working on high-profile infrastructure projects. In many cases, the workers themselves had to pay for treatment—if they could afford it at all. In others, language barriers, administrative blockages, or simple neglect led to serious delays in diagnosis and treatment.

The solution? Systematize contractor healthcare in the same way you would approach fall prevention, confined space entry, or fire safety. Make it part of your Contractor HSE Management Plan. Build it into your tender documents. Include it in your pre-qualification criteria. Ask your contractors:

- What healthcare provisions do you offer your workers?
- Do you provide insurance? Is it cashless?
- Where is the nearest clinic or hospital? How is transportation arranged?
- Do you have an onsite clinic? If not, why not?
- How are medical complaints tracked, and who reviews them?

More and more, international frameworks are catching up. ISO 45001:2018, now also calls for organizations to identify risks related not just to safety but also to worker health, physical and psychological. Clause 6.1.2, for instance, requires the assessment of risks "related to work organization, social factors at work, and working environment." This includes access to care, health surveillance, and overall welfare.

Smart organizations are going further. In Singapore, for instance, the Building and Construction Authority (BCA) mandates that all large contractors provide access to 24-hour medical services for workers, including mental health support. Their Construction Workplace Safety and Health Guidelines also require daily health monitoring for workers exposed to high-heat environments—this includes real-time tracking of body temperature and heart rate via wearable tech.

Yes, technology is playing a major role here. In Australia, the New South Wales Government partnered with contractors to pilot the use of biometric monitoring vests for construction workers in high-risk zones. These vests could detect signs of heat exhaustion and heart strain before the worker even recognized the symptoms. The results were immediate: better prevention, faster interventions, and stronger trust between management and workforce.

Even tech giants are applying similar thinking in industrial settings. At Tesla's Gigafactory in Nevada, a comprehensive on site medical system includes EMTs, nurses, and a robust digital health monitoring system. This infrastructure reduced recordable incidents and dramatically improved return-to-work rates for injured employees. While Tesla has faced its share of criticism on other labor issues, this particular move set a standard that others have since emulated.

Digitalization also enables better data. And better data leads to better decisions. Companies like Intelex, Acumen, Enablon, or EcoOnline that have invested in centralized HSE reporting platforms have been able to integrate medical tracking, incident reporting, and risk assessments into a single dashboard. This not only simplifies compliance; it also enables proactive health strategies. For example, if the system flags a pattern of gastrointestinal issues among workers in a specific accommodation block, the hygiene team can investigate water sources or food suppliers before the issue escalates.

A critical point, often overlooked: health is not just about treating illness—it's about preventing it. Nutritional access matters. Clean drinking water matters. Time off matters. Worker fatigue—especially in labor-intensive industries—is one of the most under-recognized contributors to accidents. A tired worker is an unsafe worker. A worker with no access to rest and medical attention is an accident waiting to happen.

And this is where many organizations get it wrong: they treat contractor health as an expense. It isn't. It's an investment. An investment in continuity. In quality. In morale. In resilience.

Let's return to the basics. Does your site have:

- clearly marked first-aid stations?
- trained first-aiders and medics on every shift?
- an emergency transport plan for serious cases?

- an up-to-date log of medical incidents?
- preventive health measures such as vaccinations, health checks, or screening campaigns?
- a language-accessible system for workers to report health concerns confidentially?

If the answer to any of these is no, now is the time to act.

Because health is not just a personal concern. It is a systemic one. The contractor's employee who collapses from heat exhaustion, the cleaner who works through stomach cramps because there's no sick leave, the crane operator managing blurred vision due to unmanaged diabetes—these are not isolated events. They are failures of planning, failures of policy, and failures of care. Failures no contractor management system can afford.

Accommodation

Ensuring safe and dignified accommodation for workers is not just a humanitarian responsibility—it is an operational imperative. In today's globalized contracting environment, poor living conditions can directly undermine workforce morale, safety performance, and project outcomes. And with increasing attention from media, investors, and regulators, the reputational cost of getting it wrong has never been higher.

Many forward-thinking companies are now appointing **accommodation monitors**—dedicated personnel responsible for inspecting and maintaining living quarters to a set standard. These monitors help ensure that housing provided for workers is not only compliant with national regulations but also aligns with international best practices, such as those outlined by the **ILO**.

It's easy to think of accommodation as something peripheral to safety. But housing conditions are, in many cases, a root cause of fatigue, illness, and even fatalities.

In one extreme and tragic case, workers were housed in what were described as *Portakabins*, but were in fact retrofitted **shipping containers**. Due to substandard electrical installations and poor ventilation, several workers suffered **electrocution and heatstroke**. This wasn't just a violation of dignity—it was a violation of fundamental health and safety principles.

This is not an isolated incident. A **2019 study in Addis Ababa, Ethiopia**, conducted by researchers from Bahir Dar University, revealed that **overcrowded, unhygienic dormitories** led to sleep deprivation, stress, and an increase in workplace accidents. Workers who reported poor sleeping conditions were more likely to suffer musculoskeletal injuries and reported higher instances of mental fatigue. The link between **poor housing and operational safety** was statistically significant.

When workers are tired, dehydrated, or exposed to unhygienic conditions, their reaction time, alertness, and willingness to report hazards all decline. Poor housing is not just a welfare issue—it is a **predictor of unsafe behavior** on site.

According to the ILO's **Workers' Housing Recommendation No. 115**, employers should provide or facilitate housing that guarantees:

- a **minimum floor area** per person (usually at least 3.5 to 4.5 square meters),
- a **separate bed** for each worker—*no hot-bedding*,
- clean **bedding and linen** that is changed regularly,
- **ventilation**, lighting, and protection from extreme heat or cold,
- **adequate sanitary** and **washing facilities**, with ratios such as one toilet per 6–10 workers,
- **access to clean drinking water**, laundry facilities, and pest control,
- **separate accommodation by gender**, particularly in multicultural workforces, and
- **recreation spaces** and access to common areas like canteens and restrooms.

Where temperatures exceed 30°C, **air-conditioning is not a luxury**—it is a safety measure. Prolonged heat exposure without recovery time can lead to heat stress, dehydration, and long-term kidney damage. This is not speculative—it is backed by a 2022 **ILO/WHO report**, which found a direct correlation between worker heat exposure and reduced productivity and increased accidents, particularly in construction and roadwork sectors.

For clients and main contractors, housing must be considered **in the same breath as site safety or PPE provision**. It must be included in project planning, cost estimates, and performance audits.

One leading example comes from the **Lusail City Project in Qatar**, where contractors, in partnership with Qatar Foundation and local authorities, developed the **"Workers' Welfare Standards."** These included minimum square footage per person, regulated food standards, and on site leisure spaces. Independent inspections and worker interviews were conducted, leading to a reduction in fatigue-related incidents by 30% over two years.

Another strong model is the **Dhaka Elevated Expressway** in Bangladesh. Funded partially through international agencies, the project implemented an **accommodation compliance framework** tied to contractor payment schedules. If accommodation inspections failed, progress payments were delayed.

This created strong incentives for contractors to maintain high standards—not out of charity, but as part of their operational continuity.

WHEN THE EMPLOYER MUST PROVIDE HOUSING

The ILO's Recommendation 115 says:

> It is generally not desirable for employers to provide housing for their workers directly and employers should use alternatives where possible.

This advice stems from the belief that employer-owned housing can reduce worker autonomy and sometimes lead to exploitative practices. In many urban areas, housing allowances or third-party arrangements can be more empowering.

But in **remote locations**, or in countries where migrant worker zoning regulations apply, **direct provision is unavoidable**. Two such contexts are worth discussing:

1. **Remote Infrastructure Projects:** When construction is taking place in locations with no nearby settlements—such as mining sites in Mongolia, pipeline projects in Alaska, or oil rigs off the Nigerian coast—contractors must create self-contained accommodation. These setups must include not only housing but also emergency medical facilities, fire safety plans, and mental health support.
2. **Zoning Restrictions in Gulf Countries:** In countries like the UAE, Saudi Arabia, and Oman, **municipal laws** restrict migrant worker housing from being placed in family residential areas. The intent is to prevent overcrowding, maintain social norms, and reduce community tension. Whether one agrees with the philosophy is irrelevant—the law is clear and **non-compliance can trigger legal and financial liability**.

If a client or contractor places workers in unauthorized zones and an incident occurs—particularly involving local women or children—the penalties are severe. This includes **criminal charges** for the worker, and **civil or even criminal prosecution** for the employer.

The issue of accommodation also includes **grievance mechanisms** and **worker feedback loops**. In a 2021 review by **Verité**, a non-profit labor rights organization, housing-related complaints were found to be among the top three issues raised in migrant worker helplines. Workers often lacked anonymous ways to report problems, leading to silent suffering and, ultimately, loss of trust.

Best practice now includes:

- Worker suggestion boxes and helplines.
- Regular third-party accommodation audits.
- Cultural briefings for project management teams.
- Transparent reporting of housing violations and corrective action.

One major infrastructure company in **South Africa's Eastern Cape**, building transport corridors, implemented monthly feedback forums. After receiving repeated complaints about food quality and mosquito infestations in housing units, the project team invested in better kitchen infrastructure and pest control. The investment—less than 1% of project cost—resulted in a **25% improvement in worker satisfaction scores**, which correlated with lower absenteeism and reduced disciplinary incidents.

If you're in the position of selecting or managing contractors:

- **Include housing compliance as a prequalification criterion.**
- **Audit worker accommodation facilities** as part of site visits.
- **Tie housing standards to payment milestones.**
- **Demand proof of compliance** with ILO guidelines and local housing laws.
- **Invest in training your own team** to recognize red flags in accommodation

Providing accommodation that is safe, dignified, and compliant is not only a moral imperative—it's a **strategic investment in safety, productivity, and legal compliance**. And, like all safety investments, it begins before the first worker sets foot on site.

Food

Food is never just food. It is memory, culture, dignity, and in many cases, faith. To reduce it to a commodity or a logistical problem is to miss the point entirely—and often to court risk.

It is essential to remember the strong cultural and religious influences on food. A diet that is desirable in one part of the world may be anathema in another. No-one should be required to eat what offends them for either religious or cultural reasons. In multi-ethnic labor environments, particularly in the Middle East, South Asia, and parts of Africa, this becomes a core consideration for the client's team. Halal, vegetarian, kosher, pescatarian, Jain, and other food customs must be respected—not simply as preferences but as identity markers that, if ignored, can undermine trust and morale.

This is not just about menus. It's about inclusion. And it's also about risk management. If people are working on site for more than three hours at a time, they need access to food stations on site where the food is kept safe from spoiling and where washing and toilet facilities are also available. Poor food and hygiene provisions are not only demoralizing—they can cause gastrointestinal illnesses, lost-time injuries, reduced productivity, and even outbreaks of foodborne illness that damage an entire project's reputation.

As with accommodation, any charges for food should represent a reasonable proportion of wages and should cover no more than the cost of providing the food, without any profit element. Attempting to monetize food provision often results in corner-cutting, lower quality, and resentment among workers.

A stark example of poor food management occurred during the construction of the NEOM megaproject in Saudi Arabia. In early 2024, workers at a NEOM-regulated camp staged a violent protest over the substandard quality and safety of food provision. Reports noted that food was unhygienic, nutritionally inadequate, and culturally insensitive—conditions that ultimately triggered unrest severe enough to involve law enforcement and international scrutiny.

According to the *Wall Street Journal*, utensils and debris were hurled at buildings, and workers cited repeated complaints about food quality being ignored. This incident is a powerful reminder that inadequate food provision is not merely a welfare oversight. It can become a flashpoint for protest, work disruption, reputational damage and, potentially, legal liability for failing to meet basic duty-of-care obligations.

In contrast, Singapore's approach to migrant worker dormitories—including food services—has evolved significantly, especially after the 2020 COVID-19 crisis. The government mandated standardized nutritional requirements for meals, culturally specific dietary options, and regular independent audits by third-party inspectors. Workers are also given mechanisms to provide anonymous feedback on meal quality through digital apps. By engaging both private caterers and community representatives, Singapore has created a model that balances efficiency with respect.

The result? A demonstrable reduction in food-related complaints and a documented rise in overall worker satisfaction and compliance with site health standards. The lesson is clear: systemic respect for cultural food needs is not a luxury—it is a performance enabler.

Although not directly about food, the collapse of the Rana Plaza garment factory in Bangladesh in 2013 highlighted a broader theme: welfare neglect as systemic failure. Survivors reported that the same disregard that allowed cracks to form in the building also manifested in poor hygiene and food practices in adjacent factory zones. For many multinational companies, it was a wake-up call that welfare standards needed to be holistic—and verified—not just promised in audits.

Since then, companies such as H&M and Inditex have revamped their contractor welfare policies to include spot checks on food quality, dining conditions, and worker feedback mechanisms. It's an implicit recognition that when the smallest welfare considerations are neglected, they may signal deeper cultural rot.

Poor food quality doesn't just lead to illness—it leads to errors. Numerous studies, including one from the *Journal of Occupational Health*, link inadequate nutrition to increased rates of fatigue-related incidents. For example, construction workers relying on high-sugar, low-protein meals at lunch had elevated levels of drowsiness and irritability by mid-afternoon.

Offshore operators have increasingly recognized the link between fatigue, nutrition, and safety performance. In one documented case, a contractor welfare review led to revised catering standards that prioritized sustained-energy meals and cultural suitability for a multinational workforce. Following implementation, the organization observed a reduction in fatigue-related near misses during late shifts. The lesson was clear: welfare is not peripheral to safety—it is a control measure.

So what do we need to do when it comes to food?

1. **Baseline Survey:** At project start-up, conduct a food preference and sensitivity survey. You'll be surprised what you learn.
2. **Caterer Pre-Qualification:** Include cultural competence, hygiene certification, and feedback systems as core criteria.
3. **Worker Committees:** Establish food committees with elected worker representatives to provide feedback and oversight.
4. **Audits and Sampling:** Use third-party audits. Random meal sampling isn't just good practice—it sends a message that food is a priority.
5. **Grievance Channels:** Establish anonymous reporting channels with follow-up mechanisms.
6. **Nutrition as Safety:** Make food part of the safety strategy—not just logistics.

Your client team sets the tone. If food is treated as an afterthought, that attitude filters down. But if food is framed as dignity, energy, and respect—it becomes a source of motivation, not conflict.

Because in the end, it's not just about feeding people—it's about what the food says. That they matter. That they are seen. That their culture counts. And that their health is your concern—not just their problem.

If you get that right, the ROI is not only measured in calories—but in cooperation, morale, and ultimately, safer, more productive work sites.

Mental Health

A few years ago, very few contracting companies would have considered that there was any need for them to concern themselves with this. Mental health problems were something that happened to other people, they weren't talked about and, if they were, the advice most often tendered was, "Pull yourself together." There isn't much doubt that many people wish that had not changed—but it has. And we need to recognize that migrant workers are at least as likely as any others, and possibly more likely, to be affected. They are away from home. Away from their families. Away from their wives and children. They are in a place they don't know, surrounded by people they don't know, and their freedom to do whatever they like is generally limited. They are frequently not the best educated people and may have little in the way of internal resources to keep them going when things seem hard.

Recognizing these challenges, some organizations have developed successful mental health programs tailored to the construction industry. One standout initiative is *MATES in Construction*, an Australian program that integrates mental health literacy, suicide prevention training, and on site support services. The program has shown measurable improvements in help-seeking behaviors and reduced stigma around mental health in high-risk, male-dominated environments like construction sites. Evaluations of MATES have also shown increased social support and peer engagement among workers, providing early interventions before crises escalate.

Another positive example is the approach taken by Buildcorp, an Australian commercial construction company. The company actively trains staff in mental health first-aid, openly shares personal mental health stories through leadership, and partners with professional counseling services. A notable case involved an employee, Kaine Thompson, whose early signs of distress were recognized by colleagues who quickly connected him with support. That intervention not only helped the individual recover but fostered a work culture that treats psychological safety as seriously as physical safety.

So who is watching over them? What monitoring system do you have in place to catch those who are likely to break before it happens? Unless you have trained particular people to do that monitoring job, the answer will be: none. There is no system. Until recently, that would have been the norm. It isn't the norm any longer.

A question not enough companies ask themselves is: why have these migrants moved halfway round the world to work in a place that is strange to them? The answer is: because you asked them to. It's no longer possible to say that that does not place any obligation on you. It does.

ARE YOU PREPARED FOR A MAJOR CRISIS?

It is easy to think of crises as rare or distant—a belief that persists until tragedy strikes close to home. In 2007, at the Cabin Creek Hydroelectric Plant in Colorado, five workers lost their lives and three more were injured during a recoating operation inside a long penstock tunnel. The job had been outsourced by Xcel Energy to RPI Coating, Inc.—a contractor known for its low bids and, unfortunately, its poor safety record. The tunnel, over 1,500 feet long, was a confined space, yet neither company had taken the necessary steps to treat it as such.

The crew used methyl ethyl ketone (MEK), a highly flammable solvent, without adequate ventilation or fire precautions. When a spark ignited the vapors, the resulting flash fire spread quickly through the tunnel. With only one way in and out, and no trained emergency rescue team on standby, the workers were trapped. The investigation by the US Chemical Safety Board (CSB) was damning: inadequate contractor selection, no emergency response plan, poor risk assessment, and a failure to meet even the most basic regulatory standards for confined space work.

The Cabin Creek disaster was not caused by a single misstep but by a culture of complacency and a focus on cost over safety. It remains one of the clearest modern examples of what can happen when emergency preparedness and contractor management are left to chance. It is not enough to assume the contractor knows what to do. It is the client's duty to verify, to challenge, and to demand competence.

But what about the emergencies that aren't industrial accidents? The world has changed. Wildfires, floods, hurricanes, and extreme temperatures are no longer rare events; they are regular disruptions. Climate change, whether you believe in its causes or not, is making weather more unpredictable and extreme. We now face the reality that site evacuations, supply disruptions, and weather-related downtime could occur any year—perhaps any month.

And then there are the once-in-a-generation events that have already shown us how vulnerable we are. When the COVID-19 pandemic struck in 2020, contractor operations across the globe were upended overnight. Construction sites closed, workers were furloughed or quarantined, and supply chains ground to a halt. Many firms collapsed. Others survived only because they adapted—quickly.

One example is Galliford Try, a UK-based construction company. At the onset of the pandemic, it suspended its dividend and furloughed staff to stabilize operations. With proper planning, health measures, and communication systems in place, the company managed to recover. By March 2021, Galliford Try had returned to profitability. Their resilience was not luck—it was the result of proactive risk management and an adaptable crisis plan.

Crises do not wait for convenience. They do not send invitations. When they arrive, you will either have a plan, or you will wish you had one. And that plan must include your contractors. Do they know what to do in a hurricane? Can they continue operations under health restrictions? Do they have evacuation protocols? Have they tested them?

The message is simple. Emergency preparedness is no longer optional. It is an integral part of contractor management. Not preparing is not just a failure of compliance—it is a failure of leadership. And it's always the client who pays the highest price when things go wrong.

Floods. Hurricanes. Earthquakes. Fires. Have you included them in your risk assessments? Do you know what you will do if one (or more) strikes the contract site? Is the procedure written down, and does everyone know where it is? If your answer to any of those questions is "No"—ask yourself this why not?

Before Hiring a Contractor 4

You don't write the contract after you've hired the contractor. Obviously. But you'd be surprised how many organizations approach contractor engagement backward—focusing on cost or availability first and safety, welfare, and environmental accountability much later, if at all.

I want to touch on an example I highlighted briefly in Chapter 1, In 2018, a newly installed pedestrian bridge at Florida International University collapsed unexpectedly, killing six people—including motorists driving beneath the bridge and one construction worker. The bridge had been installed using Accelerated Bridge Construction techniques, and visible structural cracks were identified in the days leading up to the failure.

Despite this, no work was stopped. Meetings were held, concerns were raised, but there was no clear contractual mechanism—or cultural willingness—for anyone to make the decisive call to halt the operation. The National Transportation Safety Board's investigation later pointed to multiple failings: poor design assumptions, lack of independent checks and, critically, fragmented communication between FIU (the client), the designer, the contractor, and third-party inspectors.

This tragedy serves as a powerful reminder that many safety issues originate long before construction begins. If early-stage contracts don't embed clear responsibilities, escalation protocols, and independent verification pathways, then even experienced teams can miss—or ignore—critical warnings.

When hiring a contractor, it's not enough to assess technical capability or cost. The structure of your relationship—how risk is owned, how decisions are challenged, and how concerns are escalated—must be formalized. The FIU incident wasn't just a structural failure; it was a failure of leadership, systems, and accountability set in motion from the very first contract draft.

If you want the right outcome, the right groundwork needs to be in place. And that means developing robust internal Contractor Health, Safety, and Environment (HSE) Management policies *before* a tender goes out, not after. It also means leveraging the expertise of every relevant department: legal, commercial, HR, procurement, operations, and HSE. Your legal team, especially, must be familiar not just with your home country's laws but also with

DOI: 10.1201/9781003754121-4

regulations in the country where the work will be executed. That often means working with specialist law firms to avoid assumptions that could prove costly later.

Let's begin with two fundamental questions:

1. Do you, as a minimum, require that your contractors' health, safety, and welfare standards match those applied to your own employees? If not, why not?
2. Do you insist on proper environmental risk consideration before work begins? If not, why not?

Consider the case of the Belo Monte Dam project in Brazil. During its planning and initial construction phases, environmental concerns raised by local communities and NGOs were brushed aside in favor of expedited timelines. Years later, the project faced widespread criticism for damaging biodiversity, displacing indigenous communities, and altering water flow patterns in the Amazon basin. Lawsuits, public protests, and international scrutiny followed, causing serious delays and reputational damage to the companies involved, not to mention strained relations with investors and financiers.

This is what happens when environmental risk is not given equal footing alongside cost and scheduling from the outset. Had these considerations been integrated into the original contractor selection and planning processes, many of these issues could have been avoided or mitigated early on. Environmental negligence is not just a sustainability issue, it becomes a business risk.

RISK ASSESSMENT: KNOWING WHAT YOU'RE GETTING INTO

We covered this concept earlier through extreme examples—like the offshore wind platform with its 200,000-word operational plan. But most contracts won't require that level of documentation. They will, however, require structured thinking.

Every contract must be preceded by a thorough hazard identification and risk assessment process. And the assessment should ask not only what hazards exist, but who is accountable and what mitigation, if any, is already in place.

A hazard is anything with the potential to cause harm, not just to employees, but to subcontractors, site visitors, and even trespassers. In the UK, for example, colliery managers were historically held personally responsible if someone got hurt on their site, even if that someone was an intruder walking a dog.

Categorizing hazards by severity helps:

- **Low**—minor injuries or operational nuisance
- **Medium**—lost time injuries, temporary shutdowns, or medical treatment
- **High**—risk to life, major disruptions, legal prosecution, or third-party impact

HOW TO PERFORM AN EFFECTIVE RISK ASSESSMENT

Engage Qualified People

Include team members who understand the technical aspects of the work. A brainstorm is not a box-ticking session. If the required expertise is not available in-house, it is essential to bring in independent specialists or consultants who possess the necessary technical or regulatory knowledge. Certain projects—particularly those involving complex engineering, environmental risk, or hazardous materials—may require niche skill sets that general HSE professionals or engineers may not fully cover.

During the construction of the Queensferry Crossing in Scotland, engineers brought in external wind and structural dynamics experts to assess how the bridge design would perform under extreme weather in the Firth of Forth. Their early input helped adjust the cable-stayed bridge design to withstand sustained wind loads, ultimately preventing delays and costly retrofitting. This forward-thinking approach was credited with the project's overall safety success and long-term resilience.

Another powerful tool that reinforces risk identification in complex environments, especially in oil and gas construction, is the Hazard and Operability Study (HAZOP). HAZOP is a structured and systematic technique used to identify potential hazards in process design or operation by examining deviations from intended performance. It is widely applied in the design and modification of chemical and process plants.

During the early planning phase of the Pearl GTL (Gas-to-Liquids) plant in Qatar, Shell implemented a series of HAZOP studies to review every critical process loop. These sessions uncovered multiple potential process safety failures, including a scenario where pressure build-up in one of the distillation units could have led to a catastrophic release. By redesigning that section of the plant before construction began, they prevented what could have been a major safety and environmental disaster.

The structured application of HAZOP allowed for engineering controls and safety interlocks to be built in at the design stage, safeguarding both workers and the facility in the long term.

Assess the Work and the Environment

Is the job being done at height? In extreme heat? With hazardous materials? Bring in Safety Data Sheets (SDSs) where relevant.

Thinking ahead at this stage can prevent incidents that only become visible when work is already underway. A potent example of this was during the Gorgon LNG Project in Australia. Due to its remote location and the sensitive marine ecosystem around Barrow Island, Chevron brought in third-party ecologists and environmental engineers early in the planning phase. Their insights led to changes in construction logistics that reduced disruption to endangered species and protected local groundwater systems, particularly by influencing how and where subcontractor camp infrastructure was built.

Had these factors been missed, the regulatory and reputational consequences would have been enormous.

Ask Hard Questions

- What could go wrong?
- What actions should follow if it does?
- What are the consequences?
- What can be done to prevent it?

Record everything in writing—even if your jurisdiction doesn't require it.

One example of early strategic planning having a measurable safety impact comes from the Sakhalin-2 oil and gas project in Russia's Far East. In its early design phase, the operator brought together internal and external experts in a multidisciplinary risk and safety review team to analyze both technical and operational risks. The team conducted comprehensive scenario-based evaluations for drilling, construction, and pipeline work, many involving extensive input on contractor roles and emergency response capabilities.

As a result of these assessments, a number of key interventions were put in place: contractor onboarding protocols were refined, high-risk interface management processes were formalized, and incident command training was mandated for supervisors. This rigorous approach resulted in significantly fewer recordable incidents during the construction phase compared to similar mega-projects in the region at the time. It also reinforced a culture of shared responsibility between client and contractors—an essential ingredient for long-term safety performance.

DOCUMENTING RISK: THE RIGHT WAY

For every identified hazard, document:

- the hazard and its potential consequences,
- the individual(s) who performed the assessment (and their qualifications),
- control measures to manage the hazard,
- department or role responsible for each control, and
- reporting pathways and response procedures.

And always remember: "Noted" is not an acceptable response.

Example: If someone trips over an unsecured cable on site, your plan should already specify who reports it, who receives that report, and what action immediately follows.

COMMON HAZARDS (BY NO MEANS EXHAUSTIVE)

- Night work without adequate lighting.
- Overhead crane operations with people below.
- Continuous exposure to high noise levels.
- Unprotected work at height.

RISK ASSESSMENTS MUST BE DYNAMIC

Construction and industrial projects evolve. So must your assessments. Reassess after:

- design changes,
- equipment substitutions, and
- scope alterations.

The Hyatt Regency Walkway Collapse is a story of small decisions with devastating consequences. During construction of the hotel in Kansas City, the original walkway support design was altered to ease fabrication. The change seemed minor—two sets of rods instead of one continuous set—but the implications were not properly reviewed. On a summer evening in 1981, with the hotel bustling during a dance event, two of the elevated walkways suddenly gave way and crashed into the lobby below killing 114 people and injuring over 200. Investigations revealed that the design change had effectively doubled the load on critical support connections, a factor that should have triggered a re-evaluation through proper engineering and risk management channels.

The tragedy highlighted the importance of procedural integrity—especially when contractors and engineers make decisions on the fly. It's a hard-learned lesson in why risk assessments must be dynamic, and why every change, no matter how small, must be evaluated not only for cost but also for safety implications affecting everyone on site, including subcontractors and support teams.

THE DANGER OF ASSUMPTIONS: GRENFELL TOWER

It is difficult to think of a more devastating case study of assumption-based failure than the **Grenfell Tower fire** in London. The 24-storey social housing block, home to many low-income and vulnerable families, was destroyed on the night of June 14, 2017, when a fire—originating from a malfunctioning fridge-freezer—raced up the building's recently installed exterior cladding. In total, **71 people died**, hundreds were injured, and a community was traumatized.

And the cause? Not arson. Not an act of terror. But a cascade of **failures driven by assumptions**.

The building had been recently refurbished as part of a local council improvement programe. Central to the refurbishment was the installation of **aluminum composite material (ACM) panels** with a polyethylene core—materials that were widely available and technically legal under the regulations at the time. But multiple parties, from the architects and contractors to the local authority and the fire service, **assumed** the materials met fire safety standards. They didn't.

Worse, some parties assumed others had verified the fire performance of the materials. Contractors assumed the client had vetted it. The client assumed the consultants had vetted it. The consultants assumed the supplier's data sheets and marketing brochures were enough.

The **Grenfell Tower Inquiry**, chaired by Sir Martin Moore-Bick, revealed a "merry-go-round of buck-passing." Everyone thought safety had been checked—by someone else. No one took ultimate responsibility to validate whether the materials met **not just the letter**, but the intent of the regulations.

The cladding turned out to be **highly flammable**. It acted as a chimney, carrying flames rapidly up the building. A series of flawed assumptions about fire stopping, escape routes, and resident vulnerability further compounded the tragedy.

The implications for contractor safety management are profound.

A phrase repeated often in project teams is: *"It's been signed off."* But by whom? With what credentials? Based on what evidence?

The Grenfell fire teaches us that it is not enough for something to have been "signed off." It is your duty—particularly during the **contractor selection and design stages**—to interrogate what that sign-off actually means.

During the pre-qualification phase, ask yourself:

- Have the materials and systems been tested *under the same conditions* they will face in your project?
- Has anyone conducted a third-party review, and is the reviewer genuinely independent?
- Is there pressure—cultural, commercial, or otherwise—to skip steps or accept industry norms without challenge?

Verification fatigue is real. Teams often default to *compliance by documentation* instead of compliance by design and inspection. "If it has the right CE mark, it must be safe." "If it meets the national fire code, we're covered." Grenfell showed that this attitude can have catastrophic consequences.

Grenfell wasn't just a story about one building—it was about a **global supply chain** failing at multiple levels. The ACM panels were manufactured by **Arconic**, a multinational supplier, and sold as meeting the required standards. But the inquiry later revealed that internal Arconic documents raised concerns about the panels' combustibility well before the fire occurred. Still, the product remained on the market, and buyers assumed its technical specifications were sufficient proof of compliance.

In contractor selection and procurement, relying on a chain of assumptions—about suppliers, materials, and delegated responsibilities—can create blind spots. One of the major lessons from Grenfell is that **risk is compounded when everyone assumes someone else has done the checking**.

A 2018 report by the UK's **National Audit Office (NAO)** pointed out that post-Grenfell safety reviews revealed over 450 other buildings across the UK with similarly dangerous cladding. In many cases, contractors and housing authorities had used the same materials, believing them to be compliant, without conducting bespoke fire safety assessments for their projects.

Though Grenfell occurred in London, its lessons have global relevance. Across the world, clients and contractors are engaging in projects with **increasingly complex design and material ecosystems**. This is especially true in countries where mega-projects—such as high-rise clusters, urban transit networks, and smart city initiatives—are rolled out under tight schedules and budget constraints.

For instance, in **India's real estate boom**, cladding, glazing, and Heating, Ventilation and Air Conditioning (HVAC) systems are often sourced from international markets, installed by subcontractors, and inspected only post-installation. A 2021 report by the **Centre for Fire, Explosive and Environment Safety (CFEES)** under Defence Research and Development Organisation (DRDO) revealed that over 70% of high-rise fires in major Indian cities involved façade components that were either non-compliant or untested under real conditions.

In **Dubai**, authorities had to amend fire safety codes after multiple high-rise fires were traced to non-fire-retardant cladding. In response, the UAE government mandated retrospective façade replacement programs—at a huge cost to property developers and building owners, some of whom had relied on written assurances from suppliers without independent validation.

Across the World, the Pattern Is the Same: Assumptions Kill

So, what does this mean for you—before you hire a contractor?

1. **Request Verification, Not Just Certification**
   If a contractor says their material or process is compliant, request third-party test data. Don't just accept CE, UL, or ASTM markings—ask to see the conditions under which the test was performed.
2. **Design for Performance, Not Just Specification**
   Ask how materials behave under stress conditions. If the contractor suggests a substitution, verify that the proposed material performs identically in context—not just on paper.
3. **Establish Accountability Chains**
   Ensure that roles and responsibilities for safety verification are contractually clear. If multiple subcontractors are involved, clarify *who* is responsible for holistic system testing and compliance.
4. **Insist on Safety in Value Engineering**
   Grenfell has taught us that design changes to save money—especially late in a project—can lead to disaster if not properly evaluated. Require all value engineering proposals to undergo safety impact reviews.
5. **Use Case Studies as Part of Tender Evaluation**
   Ask contractors to provide specific examples where they identified and rectified compliance risks before project commencement. A contractor who says "we always meet standards" might not be asking the right questions.

The Grenfell Tower fire will remain one of the most painful tragedies in the modern construction era. But if we are to honor those who died, it must become more than a symbol of failure. It must become a **compulsory lesson**—especially in the early stages of contractor engagement.

Don't assume. Verify. Then verify again. And build that expectation into every request for proposal (RFP), every material spec, and every contractor interview.

Because in safety, there is no such thing as obvious.

SAFETY CULTURE AND HUMAN ERROR

A veteran contractor removing personal protective equipment (PPE) for a moment of convenience might just be making the worst mistake of his life. And in many cases, that is exactly what happens.

In 2019, a seasoned electrician with over 20 years of experience working on a power substation project in the US was killed instantly when he bypassed insulated gloves to hand-tighten a component. He believed—wrongly—that he could avoid full contact, complete the task quickly, and return to full compliance after. He'd done it before. He wasn't being reckless in his own mind—just efficient. But a brief arc flash was enough to cause fatal injury. The investigation later showed that the correct insulated gloves were available at arm's length. He had taken them off, momentarily, because they made it harder to get a grip on a bolt.

This was not a failure of training. Nor was it a failure of rules, procedures, or documentation. It was a failure of **momentary decision-making**—an example of why **safety culture must be lived, not laminated**. And this was not a one-off. An experienced electrician in Australia made exactly the same mistake. I wouldn't doubt there have been others. Both of these men knew they shouldn't do what they were tempted to do, but there's a difference between what you know as a general principle and what lives inside you as a permanent command.

Contractor safety is often approached with the assumption that more experience equals fewer risks. But this belief can be dangerously flawed. In fact, research published by the Institution of Occupational Safety and Health (IOSH) shows that some of the most serious incidents on industrial projects involve highly experienced workers. Why? Because with experience can come overconfidence. The rules start to feel optional. They are for other people: people who haven't yet reached the same level of expertise. Familiarity breeds shortcuts. Procedures get interpreted rather than followed.

This is where the **client's role** begins long before the accident occurs. When you assess contractors before selection, it's not enough to check that they have experience or the right qualifications. You must also probe how their team manages **complacency**. Ask questions like:

- How do they reinforce procedures for experienced staff?
- Are there peer review or buddy systems in place for high-risk tasks?
- Do supervisors perform field checks based on behavior, not just paperwork?

A compelling example of how a client addressed this comes from the construction of the **New Safe Confinement** over the Chernobyl reactor in Ukraine. Managed by a consortium of global contractors, the project involved extreme radiation risks and a workforce that included many specialists with decades of field experience. The client, the European Bank for Reconstruction and Development (EBRD), mandated a *behavioral observation program* with random spot checks—not to enforce compliance per se, but to reinforce a culture of accountability. Supervisors were trained to observe micro-decisions: how workers prepared for tasks, how they handled PPE, how they acted when no one was watching. This approach reduced recordable incidents to near zero during critical installation phases, despite the complexity and danger. And it wasn't the contractors who took primary responsibility. It was the client.

Ultimately, safety culture is not measured by the presence of PPE or the sophistication of your permit system. It is measured by the decisions made **in the five seconds before the accident would have happened**—especially by those who think they know better.

And that culture must be built **before** the contractor ever sets foot on site.

PRICING AND CONTRACTOR SELECTION

You cannot and must not sacrifice safety and welfare for pricing. It's a simple sentence—so simple that it risks being dismissed as a platitude. But let's be absolutely clear: a low-cost bid that compromises quality, oversight, or accountability is **no bargain**. It's a liability.

The cost of failing to embed safety and environmental expectations into contractor selection isn't hypothetical. It's been quantified, litigated, and paid for—in lost lives, ruined reputations, operational shutdowns, and billions in penalties. And, more often than not, it stems from one decision made early in the project lifecycle: **choosing the cheapest option**.

An example comes from the **West Gate Bridge collapse** in Melbourne, Australia, in 1970. Engineers and contractors were under extreme pressure to deliver quickly and at reduced cost. A key section of the bridge buckled and collapsed during construction, killing 35 workers. Later investigations revealed that shortcuts were taken in the contractor's approach to materials and stress testing—trade-offs made to maintain budget and schedule. That cost-saving decision became one of the deadliest industrial accidents in Australian history.

Fast forward nearly 50 years to another example we briefly mentioned earlier as the lesson remained unlearned. We've already mentioned the 2013 **Rana Plaza factory collapse** in Bangladesh which killed more than 1,100 garment workers. While not a construction project in the traditional sense, the principle is identical: brands outsourced work to low-cost contractors who delivered "value" on paper but ignored structural risk, site conditions, and legal compliance. The result? A global scandal, millions in reputational damage, and widespread calls for reform in how contractor accountability is defined.

These events are often described as failures of engineering, or law, or enforcement. But at their core, they are failures of **contractor selection criteria**. The incentives were wrong from the start. The greatest damage to reputation and turnover from the Rana Plaza collapse was not suffered by the factory owners or their contractors; it was the Western retailers who had subcontracted manufacturing without checking safety who paid.

DON'T JUST ASK FOR A SAFETY PLAN—ASK FOR PROOF OF INTEGRATION

Many contractor prequalification forms include a section on safety. That's a good start. But if the safety plan is a stand-alone document and not embedded into cost, schedule, manpower planning, and subcontractor oversight, it's just paperwork. You want integration. You want demonstration.

When evaluating bids, ask:

- **How has safety been budgeted?**
- **What trade-offs have been made to reduce costs?**
- **How does the contractor ensure that subcontractors follow the same standards?**

During the **Crossrail project** in London, one of Europe's largest underground rail infrastructure efforts, safety expectations were integrated from procurement onward. The client, Transport for London, set minimum safety scoring thresholds that contractors had to meet *before* commercial bids were even opened. These scores were derived from detailed evaluations of previous safety performance, investment in workforce welfare, and ability to manage complex, multi-contractor environments. Contractors who scored low on safety were eliminated, regardless of how attractive their pricing might have been.

The result? Despite the massive scale and complexity of the project, Crossrail maintained one of the best safety records of any European transport megaproject.

The pursuit of savings in contractor selection often comes with hidden costs. Cheaper contractors may lack:

- trained HSE personnel,
- safety-qualified subcontractors,

- compliant materials or equipment, and
- systems for hazard identification, incident reporting, or escalation.

And each of those missing elements becomes a **cost** the moment something goes wrong.

This is what we mean when we say: **a cheap contractor can be the most expensive decision you make**.

Safety is only half the equation. Environmental risk, too, must be non-negotiable. And the consequences of ignoring this can be global.

Returning to the example of the **Deepwater Horizon oil spill** in 2010. British Petroleum (BP) had subcontracted rig operations to Transocean and various other service providers. In the interest of cost control and drilling speed, safety protocols were deferred, environmental assessments were rushed, and cross-contractor communication was minimal, and of course, the rest is history.

Would stronger environmental accountability at the contractor selection stage have prevented this? Quite possibly. A review by the US Chemical Safety Board cited lack of integrated oversight between BP and its contractors as a critical factor in the sequence of failures.

This isn't just an oil and gas problem. Construction projects that disregard erosion control, groundwater protection, or air quality measures often face fines, permit suspensions, or community opposition that derails progress entirely.

Lesson: just as you require technical compliance, **require environmental compliance**. From day one.

Best Practice: The "Balanced Scorecard" Approach

One of the most effective ways to embed non-negotiable criteria is to use a **weighted scorecard** for contractor selection. This approach assigns points across several categories:

- Health and Safety (25%)
- Environmental Responsibility (20%)
- Technical Capability (20%)
- Commercial Offer (20%)
- Previous Performance & References (15%)

If the bid scores poorly on HSE or environment, it should be disqualified—even if the commercial score is excellent.

Many multinationals now adopt this framework, including **HSBC for data center builds**, **Siemens for industrial installations**, and **Bechtel for infrastructure megaprojects**. In each case, scoring matrices have ensured that no bid can succeed on cost alone. The result has been lower long-term cost, stronger contractor alignment, and fewer legal or reputational issues downstream.

As the client, you hold the pen before the contract is signed. That's the moment to define expectations.

Clearly specify:

- Minimum safety standards (ISO 45001, incident rates, PPE policies, etc.).
- Environmental safeguards and local legal requirements.
- Minimum training or certification for key staff.
- Site welfare provisions (housing, sanitation, rest breaks).
- Subcontractor approval and control process.

And here's the most important part: **back it up with consequences**. If the contractor fails to meet these standards, make it enforceable. Withhold payments. Terminate work. Escalate findings.

Because if there are no consequences, then they aren't truly "requirements."

Safety and environmental responsibility must be foundational—not transactional—in contractor selection. If you start by selecting on cost and retrofitting responsibility later, you're asking for trouble.

Don't just ask how cheap the bid is. Ask:

- Is this a company I can trust to protect my workers?
- Will this team escalate concerns without fear or delay?
- Are their values aligned with mine?

If the answer is no, the bid should not proceed—no matter how attractive the price.

In **Chapter 5**, we'll explore how to practically apply this philosophy in your contractor evaluation frameworks and the bid review process. But for now, make this commitment clear: **you will not sacrifice lives or the environment to save a percentage point on the budget.** It's a promise worth making—and keeping.

Hiring a Contractor

5

FIXED COST, TIME AND MATERIALS, OR COST-PLUS?

You're the people letting the contract. So you're the people who, ultimately, are going to pay for it. What's the best way to structure the price? Best for you, best for the contractors, and best for the people working on the job.

Put it that way and it seems clear that the best price is the one that gives you a completed project on which you can make money, allows the contractors and subcontractors to make enough profit to stay in business, and provides for the health and safety of everyone employed on the project at whatever level in whatever capacity.

That's a neat summary. And it's a lot easier to say than it is to achieve.

I remember talking to someone who had negotiated a deal for the sale to the UK Ministry of Defence of computer equipment manufactured by his company. The salesman was someone I always found friendly and easy to get along with, but the Ministry's lead negotiator got up his nose in the worst possible way. Among other things, he told the salesman, "We need to see all of your internal documentation. We'll establish what your real costs are and we'll decide what margin we'll allow you to make. And it won't be high. If you're not prepared to work for a 3% margin, say so now and we'll find another company who sells what you sell."

I asked the salesman, "How did that work out?"

"We made a 50% margin. We don't normally look for anything like that kind of return, and especially not when it's taxpayer money, but I thought, 'I wouldn't talk to anyone like that, and I'm not letting you talk to me that way.' Asking to see our documentation was fine—as long as they didn't want to get their hands on confidential design information, they were entitled to see what they wanted. But we were negotiating a fixed-price contract, not cost-plus. If they'd wanted cost-plus, we'd have talked about it. And the margin—the

DOI: 10.1201/9781003754121-5

'plus'—would have been part of the negotiation. It wouldn't have been handed to us on a 'take it or leave it' basis. I have more pride in my professional abilities than that."

"So you screwed them?"

"That's not an expression I care for. We negotiated a deal satisfactory to our shareholders. And the Ministry guy got to see all the documents he asked for, and all of them were genuine."

There's a lesson to take from that story, and it's this: people don't like to be threatened. Nor do they like to be patronized. If you start from a position that says, "What we are looking for is a win/win deal from which everyone benefits," you have a better chance of securing the deal you want than if you throw your weight around. And, however expert you think your financial people are, never overlook the possibility that the other side have people who are just as good. Or better.

So: what are these three options?

FIXED COST

This is the easiest to describe, though not necessarily to negotiate. In a fixed cost contract, the parties agree before work begins exactly how much is going to be paid when the work is finished. This can go all the way down the chain, starting with a fixed cost agreement between the project owner and the main contractor and ending with a fixed cost agreement between the smallest subcontractor and someone doing a job for him. If the job ends up costing more than the agreed price, the contractor loses money. If the final cost is less than anticipated, the contractor makes an additional profit, over and above what was expected.

That sounds straightforward. And the idea has clear attractions. The contractor works out how much completing the contract is going to cost, adds in a desired profit and "contingencies" for those things that no-one can foresee but that are going to happen, and knows how much he's going to earn when the job is done. The client knows what the bill is going to be. Certainty on both sides—what an excellent thing! If only real life was like that. In fact, fixed cost contracts often turn out in practice not to be fixed at all.

A fixed cost contract is always going to represent a risk to the contractor, the risk being: if the contractor spends more to complete the job than the contract allows him to be paid, the contractor will lose money. Make a habit of entering into contracts like that and the contractor will be bankrupt. And the bigger and more complex the job, the more difficult it will be to work out exactly what it's going to cost to complete and therefore the more likely that cost will exceed revenue. Take the construction of an offshore wind farm described earlier. That is immensely complex, many contractors and subcontractors are involved, progress is subject to so many factors outside the contractor's control like weather and completely unexpected events that committing at the outset to complete the wind farm for a fixed figure may be disastrous. The strong suggestion, then, is that contracts that are truly fixed cost should be used only for fairly small jobs where what has to be done is absolutely clear.

For example, a fixed cost contract could well be suitable for a job painting the wind farm's control room entered into when the control room and its surrounding environment has already been completed and the painting contractor can clearly see the work that has to be done and the access and storage arrangements available.

A cautionary tale comes from the Olkiluoto 3 nuclear reactor project in Finland. Although not purely a construction example, it's highly instructive. This was a fixed-price turnkey contract, initially agreed at €3 billion between Areva-Siemens and the Finnish energy company TVO. Over time, however, the project experienced repeated delays, technical issues, and massive cost overruns that ballooned the final cost to more than €11 billion. To stay within their original bid, contractors were accused of rushing phases of construction, which led to quality assurance issues that had to be corrected later—at great expense. While the primary focus of reporting was on delays and financial losses, the Finnish Radiation and Nuclear Safety Authority (STUK) cited safety management concerns in how the subcontracted work was controlled and audited. With multiple subcontractors and pressure to avoid further overruns, oversight suffered. Although no catastrophic safety event occurred, the case showed how underestimating complex project costs in a fixed-price structure can create incentives for contractors to take shortcuts, undermine quality and safety assurance, and ultimately affect the integrity of the entire project. It serves as a warning for major infrastructure contracts where safety-critical work is performed under strict cost ceilings.

We are all familiar with the way huge contracts routinely suffer cost overruns: they are in the press all the time. In 2017, a study found that 64% of ongoing construction projects globally had been affected by what the study described as "cost underestimation issues." The reasons are many, but they are not what this book is about and they don't belong here. There are, though, a number of modifications to the standard fixed cost contract that have been designed to deal with the problem.

FIXED-PRICE INCENTIVE CONTRACTS

This type of contract is drawn up with a target cost and target profit. Those two added together make the target price. But, as well as the target figures, real life projects also have an actual cost. The contract includes a formula that means that if, when the contract is completed, the actual cost comes in below the target, the contractor receives a portion of the savings in addition to their profit. Conversely, if the actual cost exceeds the target, the contractor's profit is reduced proportionally. The idea is to align the contractor's financial interests with those of the client by incentivizing cost control and efficiency.

But how does this relate to health and safety?

Structured carefully, incentive-based contracts can promote proactive behavior. Contractors who know they can increase profits by completing the work efficiently are more likely to invest in planning, training, and technology upfront. When managed properly, this can lead to improved safety outcomes. However, if the incentive mechanism is overly aggressive or not coupled with robust performance oversight, it can have the opposite effect, encouraging shortcuts that compromise safety.

A positive example can be seen in the US Department of Defense's approach to some of its major infrastructure maintenance programs, such as the military base refitting projects under the US Army Corps of Engineers. In those cases, incentive contracts included explicit safety Key Performance Indicators (KPIs) as part of the incentive formula. Contractors that met or exceeded safety goals received bonuses, while those with high incident rates had profits reduced. A 2019 review by the Government Accountability Office (GAO) found that this dual focus on cost and safety drove better upfront project controls and planning, reducing the number of workplace incidents by 30% over similar past projects.

On the other hand, a cautionary tale comes from a fixed-price incentive contract used in the expansion of a chemical manufacturing facility in Texas. The contractor had an aggressive bonus structure for early delivery, but similar encouragement was not provided for safety performance. As a result, corners were cut during scaffolding erection and confined space entry procedures. The Occupational Safety and Health Administration (OSHA) later investigated a fatal accident at the site and concluded that the rush to deliver early had led to failures in contractor training and supervision. This underscores that incentive contracts, while potentially powerful, must include balanced KPIs that value safety as much as cost and time efficiency.

Ultimately, if you're considering a fixed-price incentive contract, it's vital that health, safety, and welfare provisions are not only budgeted but also embedded into the incentive structure. Incentivizing contractors to innovate, plan more effectively, and minimize risks can work in everyone's favor—but only if safety is an equal pillar in the measurement of success.

FIXED-PRICE CONTRACTS WITH ECONOMIC PRICE ADJUSTMENT

This type of fixed-price contract takes account of some things that can happen while the project is ongoing over which the contractor has no control (such as the sharp increase in material and supply chain costs seen globally during the second half of 2021). If these contingencies occur, the price can be adjusted—upward or downward—based on agreed-upon criteria.

From a health and safety standpoint, this flexibility can be vital. When contractors are locked into rigid fixed-price agreements and material or labor costs rise unexpectedly, they may attempt to preserve margins by cutting corners—sometimes at the expense of worker welfare or site safety. Economic price adjustment clauses help prevent this by providing a pathway to renegotiate pricing fairly without compromising operational standards.

A pertinent example is the $1.5 billion New Royal Adelaide Hospital project in Australia. Although not strictly an EPA contract, material cost increases and inflexible pricing mechanisms during construction created intense financial strain, leading to disputes between government and contractor. Safety regulators flagged risks associated with cost-driven staffing cuts, delayed maintenance, and poor safety oversight as contributing factors to incidents on site. Experts have since recommended that long-term complex projects like this one include economic adjustment clauses to help balance financial health with safety obligations across the supply chain.

FIXED-CEILING-PRICE CONTRACTS WITH PRICE REDETERMINATION

This type of contract comes into its own in multi-stage contracts when it is possible to fix a fair price for the first stage, but not beyond that. A ceiling above which the price cannot rise is established before work begins, and pricing can be adjusted within that limit as new stages unfold.

From a health and safety standpoint, this approach provides a level of flexibility that can be crucial. Allowing price redetermination within a capped framework gives contractors less incentive to cut corners when unexpected challenges arise during later stages. For example, if a hazardous material is discovered during the early excavation of a project, the ability to renegotiate within the ceiling allows the contractor to factor in proper disposal and personal protective equipment (PPE) upgrades, and possibly engage specialist support—all without having to absorb those costs themselves or sacrifice safety compliance.

A relevant case is the Hanford Site cleanup project in Washington State, USA. In one phase of the US Department of Energy's efforts to decontaminate this former nuclear production complex, a ceiling-price contract with redetermination provisions allowed for price adjustments as unforeseen subsurface hazards were encountered. The contractual flexibility meant that the contractor could respond safely and adequately, procuring appropriate radiation shielding and updating risk assessments without compromising worker safety to protect margins. This model helped maintain the health and welfare of thousands of on site personnel, reinforcing the value of such a contract structure in safety-critical environments.

FIRM FIXED-PRICE LEVEL-OF-EFFORT CONTRACTS

This type of contract was introduced by the US government and is concerned with the amount of work the contractor is required to perform and not with materials. A feasibility study would be an excellent example. It requires the contractor to provide a specific amount of work for a specific period, for which a specific price will be paid.

HAVE YOU INCLUDED HEALTH, SAFETY, AND WELLNESS COSTS?

Just one point—we've written a great deal so far in this book about the costs of looking after the health, safety, and well-being of people working on the project and how the benefits can be such that the costs are likely to be more than recouped. But that doesn't mean they needn't be budgeted for. In any well-run organization, when you want to do something that will require payment, the first question Financer will ask is: "Is it in the budget?" So make sure it is. If you have budgeted for an expense, and if the project has been approved and a contract signed on that basis, putting the expense through the books will be fairly simple. If you haven't, it won't.

PROJECT CREEP—WHAT HAPPENS WHEN THE PROJECT'S SCOPE CHANGES

Very few major contracts end with the precise achievement of the objectives established at the outset. Things are added, which were not included in the original costing but still have to be paid for. Things are removed because they prove impossible to achieve or the passage of time means they are no longer necessary. We're not going to go into the details here because—once again—it's outside the scope of this book, but a contract must include at the outset a procedure and set of documentation for requesting, quoting for, approving, and carrying out changes to the original contract.

TIME AND MATERIALS

In a time and materials contract, the client pays the contractor an agreed rate for the amount of time the contractor's crew and subcontractors work on the project and also pay for all materials used at cost. The drawback from the client's point of view is that there is no certainty what the ultimate price will be; from the contractor's point of view, it is that these contracts require careful administration to ensure that every minute of work is accounted for. The advantage to the contractor is that it should be impossible to lose money and, provided that the price agreed for time includes a satisfactory percentage for profit, the contractor should make money.

However, from a health and safety perspective, these contracts can present challenges, especially in high-pressure environments such as plant turnarounds or shutdowns in the oil and gas industry. These events often involve a compressed timeline and a rapid influx of external contractors—sometimes more than the project genuinely requires. Supervisors trying to avoid delay penalties may bring in surplus labor "just in case," resulting in overpopulated sites and congestion, which increases the likelihood of safety incidents.

For example, during a major refinery turnaround in the Middle East, a client authorized the engagement of over 1,200 contractors for a job initially estimated to require 800. The excess manpower led to overcrowded work zones, delays in permit processing, and lapses in effective supervision. Confined workspaces were shared between multiple teams without proper coordination, increasing the risk of incidents. In one instance, a simultaneous hot work and gas line inspection occurred in close proximity and only timely intervention by a vigilant safety officer avoided a serious incident.

This highlights the need for accurate scoping, better planning, and real-time monitoring when using time and materials contracts, especially in complex environments. Clients must strike a balance between having sufficient resources to deliver and maintaining safe, manageable working conditions.

COST-PLUS CONTRACTS

Sometimes, it is difficult to know in advance exactly what the job is going to cost because not everything in the project is settled. Some parts are fuzzy, some have not yet been fully developed in the client's mind, and some depend on external factors of which the results are not yet known. That can be when the cost-plus contract comes into its own. All of the costs are revealed to the client and this usually involves showing the client every invoice paid by the contractor and/or subcontractors. The "plus" element can be either an overall fee agreed in advance or a percentage of the total costs.

As with time and materials contracts, cost-plus contracts generally ensure that the contractor does not lose money as contractors are reimbursed for all actual costs plus a negotiated fee. However, the downside for clients is the potential for unchecked cost escalation if not properly managed. To mitigate this, clients should only enter into cost-plus arrangements with contractors who have a proven track record, preferably ones they have worked with before or who come with strong, reliable recommendations. For example, the US General Services Administration (GSA) has successfully implemented cost-plus contracts with known vendors for large-scale federal building projects by integrating strict auditing, clear deliverables, and transparent reporting procedures. These guardrails helped ensure that project costs remained justifiable while allowing the contractor flexibility to respond to scope changes. Without such controls in place, costs can quickly spiral—especially if contractors are unfamiliar, lack clear documentation standards, or can obtain benefits by expanding billable items beyond necessity. Strong oversight and mutual trust are therefore critical to the success of cost-plus contracts.

YOU THINK CHEAP IS BEST?

According to Bent Flyvbjerg, Emeritus BT Professor in Major Program Management at Oxford University's Saïd Business School, for each 100 major construction projects entered into around the world, 90 will run over budget. There are many reasons for this; one that perhaps gets less attention than it should is that the budget was never going to be adequate to get the job done properly in the first place. There is, sometimes, a tendency to go for the cheap tender even when it's clear that the bidder will struggle to complete the job for that price.

Why do people do that? We don't know. Yes, we do. It's a natural parsimony which is not, of itself, bad. A memory of a grandmother who used to say, "If you haven't got the money for it, you can't afford it." A desire to do the very best for one's company—to get them the best deal; to save money so that other projects can also be considered. And there's a less respectable version of that, which is the knowledge that we really want this project to be approved and if we let the shareholders (or the voters) know what it's really going to cost, approval is unlikely. So we pretend we can get it done for less and we find a bidder whose headline price we think will pass the scrutiny of those shareholders or voters.

And why do they, the bidders—why do they quote less then it's really going to cost? Once again, there are a number of possible reasons. One is that they've never managed a project of this size and they don't really know what the final cost will be. Another is that they haven't really understood everything that is involved and so not everything has been included in their pricing. We can sum up both of those possibilities succinctly: they're not competent. Is it really a good idea to entrust a project to a non-competent contractor?

Another reason that's probably more common is that they recognize as a kind of universal truth that all projects overrun their budget and they expect you to roll over and accept it when this one does the same. In other words, they are deceiving you. Is it really a good idea to entrust a project to a deceitful contractor?

And then there's a trend that has become increasingly common in international infrastructure projects: contractors, often backed by significant state or institutional support, submit extremely low bids to secure major projects. These contractors effectively "buy the business" by offering prices that appear too good to pass up. Many times, the pricing is artificially low because of indirect financial support or strategic long-term positioning. From a purely financial perspective, clients may be tempted to accept such bids, reasoning that if a contractor is willing to assume the financial risk, why not let them? There is, on the surface, a logic to that argument.

Except that, in a number of those projects, the contractor recovered some of the costs of buying the business by skimping. One tender for a road project in the Caribbean was just over 50% of the next lowest bid. That tender won. The road was built. It wasn't built to specification, though that did not become clear until a road-marking contractor found that, on curves, he was laying the edge line on bare soil—the road was simply not as wide as it was supposed to be. And then, as traffic passed, excessive wear on the road revealed that the asphalt was not as deep as it was supposed to be, either.

Should the supervising project manager have picked those faults up at the time? Of course. Why didn't that happen? I'm sorry, I'm not commenting on that.

So does that mean that the safe course is to accept the highest offer? Of course it doesn't. But what you do need to do is to examine very carefully each contractor's record and to compare in great detail exactly what each contractor is proposing to deliver. And, speaking of records.

ORGANIZATIONAL CAPABILITY PROFILES

The Organizational Capability Profile (OCP), frequently presented as a chart, describes the skills, knowledge, and resources that the company bidding for your business claims will enable it to meet your requirements in all aspects. Ask to see it, if it is not included in the bid. The OCP and supporting technical documentation should be written in clear and understandable language. If anything seems in the least flowery or—worse—imprecise, ask for a written explanation of what it means.

WHAT IS THE CONTRACTOR'S HSE RECORD?

It's too easy for Health, Safety, and Environment (HSE) to become something that is talked about more than it's practiced. If your Invitation to Tender told bidders that health, safety, and well-being would be factors in awarding the contract (and, if it didn't, it should have), then every bidder who expects to be taken seriously should have included relevant material. It isn't enough just to read the bidder's assurances that these matters receive their full attention. Anyone can say that. Do they do what they say they're going to do when they are out there on site?

Begin by asking for copies of the contractor's internal HSE processes and documentation. Examine them. Carefully and in detail. Do they describe the kind of process that we've been talking about throughout this book so far? If they don't, you need to identify the areas in which they fall short, take these up with the contractor, and get them written into the bid.

Something else the bidder should make clear (and you should ask for if it isn't there) is documented proposals on how the contractor will handle on site waste and housekeeping. These are essential parts of the HSE hands-on nitty-gritty that are often missed out when the bidder is talking the talk but doesn't have too much experience of walking the walk.

Next, you should not only ask for references—firms for whom the contractor has successfully carried out projects in the past—but also look for the opportunity to visit one of the contractor's present working sites (ideally, unannounced). While you're there, you'll be able to confirm that the contractor is delivering for that client what the tender promises to deliver for you. If that isn't happening, you need explanations, because what you're seeing tells you whether best practice is—or is not—normal for this contractor.

And make the inspection thorough. Where are the food and water stations? How about the qualified first-aiders, their nearness to the places where people are working, and the facilities and supplies available there? If the contractor is providing accommodation, have you inspected it? Does it have everything we listed as necessary in the previous chapter? If the site is in a location with extreme weather, what is being done to protect workers from excessive heat or other risks?

CLIENT-CONTRACTOR POWER DYNAMICS

One of the most underexplored yet critical risks in contractor management isn't technical failure—it's power imbalance. When clients exert dominance—whether intentionally or tacitly—it creates a dynamic where contractors may feel too pushed or too ignored. Either extreme can erode safety culture and lead to breakdown.

A recent study in South Africa examining over 100 construction projects concluded that limited client involvement in safety management was a key predictor of incidents, even when compliance frameworks existed on paper. The message is clear: health and safety responsibilities cannot be outsourced in name alone. A contract does not replace culture.

This power imbalance often plays out in subtle ways. It may show in how risk assessments are reviewed, how communication flows during work execution, or in how freely contractors feel they can raise concerns. In one peer-reviewed study involving over 500 workers in China, researchers found that project owners who demonstrated clear, structured involvement in safety matters created a stronger safety culture across the site. When clients set expectations and coordinated efforts across contractors, safety performance improved. However, in environments with high "power distance"—that is, where one party dominates and the other is passive—contractors became disengaged, withheld concerns, or reverted to minimum compliance. The researchers concluded that client tone directly shapes contractor behavior.

Portugal's 2021 construction safety review went a step further, showing that when clients held all the authority but little operational presence, safety culture deteriorated. Contractors in these settings often saw HSE requirements as "box-ticking" rather than values to be lived. In contrast, projects with shared governance models—where clients invited contractor reps into safety committees or risk review boards—saw more reporting of near-misses and better response to evolving hazards.

Closer to everyday reality, an American study published in the Journal of Safety Research examined how subcontractors perceived general contractor safety climates. Interestingly, smaller contractors consistently believed that their internal safety practices were stronger than the expectations or environments created by their clients. Even with rigorous prequalification systems, this disconnect persisted. It tells us something important: technical compliance doesn't always equal trust or belief in the system. Especially when the system feels distant or disinterested.

So, what can be done? First, client leaders must model the behaviors they wish to see. This goes beyond issuing policies or conducting the occasional site visit. It involves showing up, asking the right questions, listening without judgment, and acting without delay. Second, shared ownership of safety must be built into the structure of the project—through joint reviews, collaborative planning, and open communication. Third, incentive systems should reward not only performance outcomes (like injury rates) but also behaviors that build a strong safety culture: such as reporting, coaching, and leadership.

This doesn't mean diluting authority. Clients must remain accountable. But they should also be aware that their authority creates a tone—and that tone can empower or intimidate, inspire or silence. Great contractor management understands that control doesn't come from paper contracts, but from trust, clarity, and shared purpose.

As you reflect on your own organization, ask: are we visible to our contractors? Do they see us as collaborators or as silent overseers? Because in high-risk environments, silence is not neutral—it's dangerous.

SAFETY CULTURE MATURITY

When managing contractors, it's tempting to focus solely on procedures, systems, and compliance checklists. But in reality, the success—or failure—of contractor safety often comes down to one thing: culture. Not just what people say about safety, but how they act when no one is watching. Understanding where an organization sits on the spectrum of safety culture maturity can provide powerful insights into what needs to change, and how.

One of the most widely recognized tools in this area is the **Safety Culture Maturity Model (SCMM)**—a framework that helps organizations assess the behavioral and cultural aspects of safety across five broad stages. While different sectors and institutions may apply the concept in various ways, a particularly influential version was developed by Hudson (2001) and popularized through Shell and later adapted by many organizations including DuPont (via the Bradley Curve).

These models generally define the stages as:

1. **Pathological:** "Who cares as long as we're not caught."
2. **Reactive:** "We do something when there's an incident."
3. **Calculative/Compliant:** "We have systems in place to manage all hazards."
4. **Proactive:** "We work on problems before they happen."
5. **Generative/Resilient:** "HSE is how we do business around here."

The terminology may vary slightly—so me replace "pathological" with "vulnerable," or "calculative" with "compliant"—but the idea remains the same: organizations evolve in how they perceive and act on risk. The further along the maturity curve, the more safety is embedded in day-to-day decisions, and the more empowered frontline teams become.

A **reactive HSE** model is better, but still not good; contractors operating in this way have the knowledge and facilities to deal with problems, but not until they arise. Risk assessment and control are both retrospective. (An accident has happened—what was the risk?) Basic safety training is in place, but there is little motivation and it is ignored more often than it is followed. When accidents happen, they are investigated but the investigation focuses on human error rather than the environmental factors that play a part in most accidents.

A **compliant HSE** model is one that ticks all the boxes; it complies with all regulations applicable to the site. That would be the perfect solution if the regulations covered every possible problem and if the typical contractor's corporate culture was safety-oriented. Unfortunately, they don't and it isn't. Still, this is better than either of the previous two models.

Contractors with a **proactive HSE** model are looking for trouble and seek to deal with it before it happens. Now we are really talking HSE—this model has a great deal going for it and contractors with proactive HSE are likely to anticipate problems before they arise and fix them before they happen. Risk assessment and control happen in advance. There is HSE documentation, everyone knows what and where it is, and a culture exists in which the documentation is respected. Audits are integrated and appraisals include an assessment of safety performance. When accidents occur, investigation seeks to identify the cause or causes rather than to apportion blame.

And then we have what should be the gold standard of HSE models: the **resilient HSE** model. HSE is a fundamental and acknowledged part of all activities, and so are risk assessment and control. The focus is on eliminating risk, wherever that is possible. When accidents do occur, investigation looks at every possible cause from every angle. A pro-safety culture is integrated into the company.

Most clients assume their contractors are somewhere around the compliant or proactive level, but in reality, they may be operating in a reactive or even pathological state. Without a clear understanding of cultural maturity, contractor selection becomes a gamble. Conversely, when maturity is measured and understood, contractors can be supported to grow—and clients can design contracts and oversight processes that match.

One example of a contractor organization applying a safety maturity model is **Wood Group**, a global engineering and consultancy firm operating in the oil, gas, and renewables sectors. Faced with challenges in subcontractor oversight on high-risk offshore projects, Wood implemented a culture maturity assessment across several of its divisions. The findings revealed a strong calculative culture, good systems, but limited engagement from frontline staff.

Wood responded by shifting leadership training from compliance enforcement to values-based engagement. Supervisors were coached to conduct daily safety dialogues, ask better questions, and involve workers in shaping safe work procedures. Over time, they observed a 35% increase in near-miss reporting and a significant reduction in repeat incidents. By understanding their cultural position, they were able to evolve it.

Network Rail, the UK's rail infrastructure operator, introduced a cultural maturity assessment across both internal and contractor teams as part of its "Safety Step Up" initiative. The organization recognized that, while internal staff operated at a proactive level, many contractors were still reactive—waiting for incidents before acting.

Using independent assessments, including surveys, focus groups, and behavioral observations, they mapped the maturity levels and tailored interventions accordingly. Lower-maturity contractors received more structured supervision and embedded client reps. Higher-maturity partners were empowered to lead their own HSE initiatives. The result was a consistent year-on-year improvement in safety performance across the contractor base.

Where many organizations go wrong is assuming that maturity assessments are just another audit. In reality, these tools are most powerful when they're used to start conversations—not end them. For example, asking a contractor to self-assess their culture against maturity stages opens a space for honesty, reflection, and targeted support.

Moreover, clients can use these assessments to design tiered assurance models. A contractor operating at a calculative level may require detailed checklists, closeout reports, and task-specific Permit to Work's (PTW). A proactive contractor, on the other hand, may benefit more from coaching, performance dashboards, and peer-to-peer learning sessions. It's not about loosening the reins—it's about tailoring oversight to the level of cultural readiness.

While we will be discussing more about ESG (Environment, Social, Governance) in a later chapter, many organizations are evolving their contractor management models to align with broader Environmental, Social, and Governance (ESG) principles. Culture maturity fits neatly into this shift. Organizations that score high on governance and workforce welfare often demonstrate generative safety cultures—where contractors are treated as partners, not liabilities.

In 2023, a multinational energy company incorporated cultural maturity scoring into its ESG supply chain audits. Suppliers were evaluated not only on lagging indicators like injury rates, but on behavioral indicators—reporting culture, leadership engagement, worker involvement. The initiative uncovered significant maturity gaps in two key suppliers, prompting collaborative workshops that led to long-term partnerships and shared learning.

For contractor management, the lesson is clear: if you don't know where you are culturally, you can't chart a path forward. Safety maturity models offer a language and a lens to assess, engage, and evolve. More importantly, they remind us that culture isn't something you write into a contract. It's something you build through leadership, listening, and trust.

6 Monitoring

So here we are. You've called for tenders. You've examined them all and invited those who submitted the three or four best to make their pitch. You've taken up references, talked to people you know at clients they've worked for, and made unscheduled visits to their current sites to see how well they really do what they say they do. You've chosen the successful bidder, agreed the terms and signed the contract. So that's it? Now you can hand over the site and say, "Over to you?"

No, I'm afraid not. You can certainly hand over the site but overall responsibility still lies with you. You have to stay on top of the job and that means checking that the contractor is delivering what the tender promised and the contract demanded.

This is not, in Health, Safety, and Environment (HSE) terms, about noticeboards saying how many days a site has worked without accident or injury. It's partly about bringing down the number of incidents by proactively managing the contractor's HSE, and we'll deal with that first—but it's also about your own organization and how you set yourselves up to succeed.

In September 2015, during preparations for the Hajj pilgrimage, a massive crawler crane operated by the Saudi Binladin Group collapsed into the Grand Mosque in Mecca, killing 111 people and injuring nearly 400. The crane, one of many used in the expansion project, toppled during high winds and a thunderstorm.

An investigation by Saudi authorities found that the crane had not been properly secured, and the contractor had failed to follow manufacturer guidelines for wind-based operation limits. Despite weather warnings, the crane's boom had been left elevated—exposing it to wind loads far beyond safe thresholds. The official report cited gross negligence by the contractor.

From a contractor management perspective, the Mecca incident is an emblematic case of what can happen when operational risk control is delegated without verification. Although the Binladin Group was highly experienced and deeply connected to major state projects, assumptions about their internal controls replaced active oversight and monitoring. The result was

 DOI: 10.1201/9781003754121-6

catastrophic—not only in terms of loss of life, but also in reputational and operational terms. The company was banned from new public contracts for a period and faced lawsuits and widespread scrutiny.

And in a similar example, in July 1999, during the construction of Milwaukee's Miller Park baseball stadium, one of the largest cranes in the world—nicknamed "Big Blue"—collapsed while lifting a roof section weighing over 450 tons. The incident killed three ironworkers and destroyed the crane itself.

The investigation revealed a critical lack of alignment between contractor planning and environmental monitoring. Wind speeds exceeded safe operational limits for the lift, but the crew pressed ahead—either unaware or under pressure to stay on schedule. Key safety protocols, including windload assessments and emergency procedures, were either bypassed or inadequately communicated.

From a contractor management standpoint, this incident illustrates the dangers of assuming that specialist contractors always have it under control. Lifting operations involving high-risk equipment require not just technical approvals but real-time client oversight, clear stop-work authority, and contingency plans triggered by environmental variables.

For organizations managing contractors, the lesson is clear: risk monitoring doesn't end once the method statement is signed off. It must extend into the live operation, with clients and contractors both actively engaged in managing dynamic risk factors—especially when lives are hanging from a hook.

THE IMPORTANCE OF MONITORING: OUTSOURCING IS NOT ENOUGH

In Chapter 2, we mentioned the contractor who built houses on what had previously been an artillery training ground. The work was regularly interrupted when unexploded shells were turned up by diggers. There was never any chance of keeping it secret, because each time it happened the police knocked on nearby doors and told residents to stay indoors and keep away from windows. One result of that was that the story of the shells was told to anyone known to be considering buying one of the houses. Clearly an undesirable result if you happen to be tasked with selling the houses.

The possibility that a one-time artillery training ground might contain buried unexploded shells could have been foreseen. In fact, it was foreseen. A survey had been outsourced. And that is the lesson for this chapter from that incident. You've outsourced something—a survey, safety training, first-aid provision, accommodation management—whatever. Chances are, you'll find yourself outsourcing a number of things. The fact that you've outsourced them doesn't mean you can now ignore them. It's your contract. If one or more of the parties to whom you have outsourced responsibilities fails to deliver, it will be you who bears the cost. So monitoring what's happening on the ground falls to you. You have to make sure that what has been contracted for is being delivered according to the contract's terms. Lagging and leading indicators aren't the whole answer, but they can certainly help.

LAGGING INDICATORS

Lagging indicators, also known as historic, past, or reactive indicators, deal with the outcomes of actions taken in the past. There was an incident, an accident, an illness. What did you do? What did another party do, if another party was responsible? And why does it matter? (That is the easiest question of all to answer: it matters because, if you took no action, or acted but didn't measure the outcome, you have no way of knowing how you are performing against your HSE targets. It also matters because the aim should always be to improve performance and, as with anything you're involved in, if you can't measure it, you can't improve it).

So which indicators should you monitor? At the very least, these:

- How many events have taken place that can be regarded as connected with HSE? That is, how many accidents? How many incidents? How many cases of illness?
- How many near misses?
- How many days were lost as a result of the above?

Then, you need to be monitoring those outcomes in more detail:

- What impact did the days lost figure have on your business performance? And business performance isn't just about the bottom line, though you will want to see the effect on your costs and profit; any damage to your reputation should also be monitored and all facts provided to your Press Department and PR people so that they can deal with the damage.
- Rank the near misses, incidents and accidents by trade. Which trade had the most? Which the fewest? Why?

- What type of incident, accident, or injury occurred most frequently? What investigation was carried out into the causes? What actions have been taken to reduce the frequency?
- How many actions resulting from the above are outstanding, and for how long? If it's more than a few days—why?

But here's where it becomes complicated. Lagging indicators are tangible—they're quantifiable, easy to count, and often required for regulatory or legal compliance. They provide the hard evidence needed to understand what went wrong and when. However, their appeal—clarity, finality, and countability—is also their weakness when they become the focus of performance incentives.

Lagging indicators have value but they should not be set as primary goals or used directly as performance incentives. For instance, offering bonuses for achieving zero Total Recordable Incident Rate (TRIR) may seem like a way to encourage safety, but numerous studies and industry feedback have shown this can backfire. Employees and contractors under pressure to meet a "zero" goal may underreport incidents or, worse, hide them entirely. This undermines the intent of safety programs and distorts data that is crucial for understanding real risks. Additionally, tying incentives to incident-free records can shift the organizational focus from genuine concern for colleagues' well-being to frustration over lost bonuses.

One study from the National Safety Council (NSC) warned that overly punitive or reward-based lagging metrics can hinder open reporting culture and lead to a culture of silence. A well-known example is a 2009 case study involving a US energy company where bonus structures were tied directly to zero-incident targets. Investigations revealed that several minor incidents had been deliberately concealed by both contractors and supervisors to avoid the loss of financial incentives. This not only placed individuals at greater risk but also prevented systemic hazards from being addressed.

Going back to the case of British Petroleum's (BP) Texas City Refinery explosion in 2005. Before the explosion, BP was reporting low injury and incident rates—metrics that gave a false sense of safety performance. Internal investigations later revealed that the company's intense focus on personal injury rates masked deeper process safety issues, such as mechanical integrity and risk management failures. By focusing on lagging indicators like TRIR while ignoring leading indicators such as audit results, near misses, and maintenance schedules, BP's leadership missed critical early warnings.

Another pertinent example comes from the healthcare industry, where hospitals in the United Kingdom began tying staff bonuses to the reduction of patient incidents. A 2016 report by the UK's National Health Service (NHS) Confederation concluded that such targets led to a sharp decline in reporting, not because incidents were decreasing, but because frontline staff became reluctant to report errors that could impact their teams' performance bonuses. The subsequent loss of transparency and the delay in addressing actual systemic issues resulted in avoidable harm to patients. In short, the metric became the mission.

Even well-intentioned systems can go wrong. A construction firm in Southeast Asia tied a monthly safety incentive program to having no lost-time injuries (LTIs). At face value, this encouraged caution on site. But internal audits later found that several injuries were being reclassified as "first aid" to avoid triggering LTI status. Medical records were adjusted, and in one case, a worker with a fractured wrist was reassigned to paperwork rather than recorded as having lost time. When the true safety culture was reviewed during an external audit, it became clear that the incentive system had driven behaviors that undermined the very purpose of incident tracking.

The message here is not that lagging indicators are useless. Far from it. They are valuable for trend analysis, regulatory reporting, and for conducting root cause investigations. But they are tools for reflection, not prediction. They tell us what went wrong, not what will go wrong.

Contractor management teams should recognize that excessive reliance on lagging indicators can erode trust. When subcontractors feel that admitting to a minor slip-up could cost them a contract renewal or affect their standing on a client's performance scorecard, the incentive is to stay silent. In that silence, danger grows.

The better alternative is to focus on building a safety culture that rewards transparency, encourages near-miss reporting, and values proactive risk identification. Some companies now use balanced scorecards that include both lagging and leading indicators. For example, tracking participation in toolbox talks, the number of hazards reported, or the completion rate of safety training modules are leading indicators that help predict future performance.

Organizations such as DuPont, ExxonMobil, and Shell have long recognized the power of leading indicators in contractor performance evaluation. DuPont's STOP program, for instance, emphasizes behavioral observations and engagement as primary metrics. These aren't used to penalize but to coach, reinforce, and improve. Likewise, Shell's contractor safety initiatives promote the early reporting of unsafe acts and conditions, without linking them to bonuses or penalties.

The principle is this: safety must be a value, not a target. Once it becomes a target with monetary implications, it becomes vulnerable to manipulation.

If your contractor safety program relies heavily on lagging metrics, ask yourself: are you really measuring safety, or are you measuring silence? True safety comes from openness, from learning, and from systems designed not just to react to injury but to prevent it.

Lagging indicators should form part of a reflective process—used in post-incident reviews, shared lessons learned, and long-term performance analysis. But as a motivational lever or key performance incentive? That's not safety leadership. That's scorekeeping.

In the next section, we will look at how to define and implement meaningful leading indicators that drive authentic safety engagement and enhance contractor performance in real-world settings.

LEADING INDICATORS

Leading indicators are more difficult to measure than lagging indicators for the simple reason that they measure things that haven't happened yet. But that is also what makes them powerful. They allow us to gauge our performance *before* something goes wrong. Leading indicators tell us what is happening now—and put that way, it must be clear that we need to know what they say.

To understand what leading indicators are, go back to that last example of a lagging indicator—the number of actions that haven't been completed. That is backward-looking. But if we instead measure how many of those actions were closed within a specific timeframe—say, within 72 hours—we are now using a leading metric. We are tracking the system's responsiveness and discipline. That is a proactive measure.

When we monitor the performance of a project, we want to know whether we are on time and within budget. So, the first two leading indicators are fairly obvious:

- You have Gantt charts and similar visual ways of monitoring progress in each part of the project.
- How many milestones are being completed ahead of time?
- How many milestones are being completed on time?
- How many milestones are behind time?
- For the latter question, have actions been put in place to bring the delayed milestones back in line with planned completion? Are they being performed, and are they working?

You also have budgets for every part of the project:

- Where are you under budget?
- Where are you on budget?

- Where are you over budget?
- For the latter question, have actions been put in place to bring the over-budget items back into budget? Are they being performed, and are they working?

Now, since we are concerned in this book with HSE matters, let's explore how leading indicators are applied to safety performance, particularly when managing contractors:

- What percentage of corrective actions have been closed out from previous incidents?
- How many safety observations have been reported by contractors?
- How many toolbox talks have been completed per shift, per team?
- What percentage of safety inspections identified zero non-conformances?
- How many safety walkthroughs have the management team completed?

These metrics, while indirect, are deeply telling. They don't just measure results; they measure *engagement.* They indicate whether the system is functioning, whether supervisors are present and proactive, and whether the frontline workforce feels empowered to participate in safety.

Crucially, leading indicators can be positively leveraged to enhance proactive safety behaviors across teams. Rather than relying solely on lagging indicators, successful organizations are increasingly aligning contractor incentives to leading indicators. This creates a healthier, more engaged safety culture. Examples from across industries highlight this trend.

A peer-reviewed case study published in the journal *Safety Science* (Zohar & Polachek, 2014) explored the implementation of leading indicators in a large industrial manufacturing project. The company integrated proactive measures such as near-miss reporting rates, frequency of safety walkthroughs, and training completion into their contractor Key Performance Indicators (KPIs). These metrics were then linked to performance reviews—not to punish, but to encourage continuous improvement. The result? Incident rates dropped by more than 30%, and the workforce reported higher levels of safety ownership and transparency.

In another example, the construction giant Skanska implemented a system in the UK where contractors were scored monthly on leading indicators such as timely closeout of safety actions, attendance at pre-start briefings, and the frequency of "You See It, You Sort It" observations submitted by team members. According to Skanska's public HSE reporting, sites with the highest leading indicator scores saw fewer injuries and greater compliance during audits. More importantly, workers on these sites reported feeling more respected and involved in day-to-day risk management.

Oil and gas major BP also shifted focus to leading indicators after the Deepwater Horizon disaster, which had exposed fatal cultural and process weaknesses. The company's updated safety framework now includes contractor compliance rates with process safety drills, the frequency of hazard hunts conducted per platform, and leadership engagement scores derived from site visits. These metrics are reported monthly and influence decisions on contractor renewals and awards. The new approach—outlined in the *BP Sustainability Report*—has led to what the company terms "safer and more stable field operations."

The US Department of Energy's Hanford Site Cleanup Project adopted a comprehensive behavior-based leading indicator program to improve contractor safety engagement. Instead of focusing only on incident rates, contractors were evaluated on their performance in areas like job safety analysis participation, peer-to-peer safety observations, and safety innovation submissions. Within two years of implementation, the project's lost-time incident rate decreased by 45%, and near-miss reporting increased by 200%, illustrating not a rise in risk but a rise in awareness.

Similarly, a 2021 study by the National Safety Council (NSC) profiled a logistics firm that overhauled its contractor safety program to focus on leading metrics such as driver fatigue assessments, vehicle inspection compliance rates, and completion of daily pre-trip safety checks. This was paired with an incentive program where contractors could earn bonus points for exceeding safety engagement thresholds. According to NSC's findings, turnover dropped, incident rates halved, and contractor satisfaction rose—transforming the safety culture from one of avoidance to one of participation.

What do all these examples have in common? They reward the right behaviors. Instead of rewarding silence (which is often what lagging metrics do), they reward contribution. They align incentives not with "nothing bad happening," but with "something good being done."

That's the heart of the matter.

When we tie incentives to reporting a hazard, or leading a toolbox talk, or closing a finding ahead of schedule, we communicate what we truly value. And when people believe they are being measured for their involvement—not just their compliance—they become participants in the safety system, not just subjects of it.

However, using leading indicators effectively in contractor management does require structure. It cannot be a scattergun approach. Here are best practices to keep in mind:

1. **Set Clear Definitions**: Not all leading indicators are created equal. Choose metrics that are specific, observable, and relevant to your work environment. "Quality of toolbox talks" is vague. "Percentage of shifts that begin with a documented safety briefing" is measurable.
2. **Link to Risk**: Choose indicators that reflect the top risks in your operation. For example, if working at height is a critical hazard, track the number of pre-task height safety assessments completed.
3. **Include Contractors in Development**: Don't impose metrics without consultation. Involve contractors in identifying and agreeing on indicators. This builds ownership and relevance.
4. **Use Data for Coaching, Not Punishment**: Metrics are not baseball bats. Use them to start conversations, offer support, and celebrate improvements. If someone is falling short, ask: what's the barrier? What's the fix?
5. **Communicate Results Widely**: Share dashboards, trends, and lessons learned openly. Recognize high-performing teams and give context to areas needing improvement.
6. **Review and Evolve**: What works this year may not work next year. Reassess your indicators as risks, technologies, and workforce dynamics evolve.

When done well, leading indicators are not just a measurement tool—they are a mindset. They promote anticipation over reaction, responsibility over blame, and culture over control.

So let us return to the central theme of this chapter: safety must be embedded *before* contractor selection. This includes the indicators you choose to manage performance. If you wait until incidents occur to evaluate a contractor, it's too late. Instead, align with contractors who are already tracking and acting on leading indicators. Make this part of your prequalification process. Ask for their data. Ask for their learnings. And ask them how they support their own teams to act proactively.

It is not about being perfect—it's about being prepared. That's the lesson of every great safety organization. They do not assume safety; they build it every day, from the ground up.

And they use leading indicators to light the path forward.

KEY PERFORMANCE INDICATORS (KPIS)

Any discussion of leading indicators inevitably leads us into the territory of KPIs. This isn't a coincidence. It's because leading indicators—the things that show what is happening right now, before harm occurs—are the lifeblood of intelligent, proactive KPIs. And when it comes to contractor management, getting KPIs right is not only a technical issue. It is a cultural one.

We set KPIs to send signals. The targets we select—and what we choose to reward or penalize—tell people what really matters around here. If the message sent by our KPIs is that budget adherence is more important than worker safety, don't be surprised when corners are cut. If we signal that "zero accidents" is more important than learning from near misses, then reporting stops and the problems grow silently, underground.

The principle of KPIs is deceptively simple: set a target, measure progress, and act accordingly. But there's a trap here. When designed carelessly or narrowly, KPIs can push people toward the appearance of success while actively eroding the safety culture we're trying to build.

The Chemical Safety Board (CSB) makes this plain. In the aftermath of the **BP Texas City refinery explosion** discussed in earlier chapters, investigators found that KPI dashboards painted a glowing picture of safety performance, based largely on lagging indicators like recordable injuries. Meanwhile, systemic process safety risks were growing unchecked. Lagging indicators alone had given leaders a false sense of security—until it was too late.

The solution isn't to abandon measurement. It's to measure what actually matters—and what can be acted upon *before* things go wrong.

Organizations with mature contractor management systems take a dual approach. They use lagging indicators to learn from the past, but they focus their KPIs—especially those tied to incentives—on leading indicators. The best in class do five things:

1. **Balance leading and lagging indicators, but skew toward the leading.**
2. **Avoid tying financial rewards to "zero incident" targets.**
3. **Incentivize behaviors, not just outcomes.**
4. **Focus on transparency and learning, not blame.**
5. **Tailor KPIs by role, trade, and risk profile.**

Let's explore how this works in practice.

Shell's "Goal Zero" initiative is often misinterpreted as a focus on zero incidents—a classic lagging indicator target. But the program's success lies in its layered approach to KPIs.

Shell tracks traditional lagging metrics like Lost Time Injury Frequency (LTIF), but also deploys an extensive set of leading indicators, particularly in contractor-heavy projects. These include:

- Percentage of contractor personnel completing safety inductions.
- Completion rates of daily toolbox talks.
- Frequency and quality of leadership safety walkabouts.
- Timeliness of action closure from safety observations.

In one internal review of a major offshore project, Shell found a direct correlation between the frequency of quality leadership engagements (a leading indicator) and reduction in near-miss events. The message: safety leadership is not symbolic—it's measurable, and it matters.

The Construction Industry Institute (CII) in the US has conducted extensive research into contractor safety performance. In a multi-project study published in 2016, CII found that projects with high leading indicator performance—defined by proactive metrics like hazard identification reports, field safety inspections, and training completion—had 73% fewer recordable injuries than those relying solely on reactive metrics.

Key takeaways from the study included:

- Incentives based on observation and hazard reporting drove stronger engagement than those tied to incident-free milestones.
- Projects that recognized *quality* of safety interactions—not just quantity—had stronger contractor buy-in.

In short, leading KPIs didn't just improve numbers—they changed behavior.

The Qatar Rail program, a multibillion-dollar infrastructure development that was voted as the World Wide Operator of the Year by the Global Light Rail Awards in 2024, implemented a contractor performance measurement system that integrated both types of indicators. But the breakthrough came when they transitioned from tracking only reactive metrics (e.g., recordables and fatalities) to proactively measuring:

- Near-miss reporting frequency (per 100,000 work hours).
- Number of positive safety interventions per week.
- Safety meeting attendance rates.
- Percentage of workforce completing behavioral safety training.

By rewarding contractors who consistently reported near misses—even without incidents—Qatar Rail helped foster a no-blame culture that encouraged openness and early risk identification. This was critical given the multinational contractor environment and complex interfaces between firms.

UK-based construction and engineering firm Balfour Beatty integrates "Visible Felt Leadership" as a key leading indicator. This measures the frequency and effectiveness of senior leader safety engagements—site visits, walk-arounds, and one-on-one conversations with workers.

The International Association of Oil & Gas Producers (IOGP) has published comprehensive guidelines on contractor HSE performance indicators. Among the recommended leading KPIs are:

- Number of HSE training hours per employee.
- Frequency of documented inspections.
- Ratio of reported hazards to total workforce size.
- % of actions closed within timeframes.

The IOGP emphasizes that the use of standardized leading indicators across global contractors allows benchmarking, trend analysis, and improved oversight—particularly in complex, high-risk projects.

But what should contractor KPIs look like?

While every organization and project is different, there are core categories that should form part of a contractor KPI suite. Consider including KPIs in the following areas:

1. **Training and Competency**
    - Percentage of workforce with valid certifications.
    - Percentage of contractor staff completing client-mandated training.
    - Audit results of on-the-job competency.
2. **Hazard Identification and Resolution**
    - Number of hazards reported per 100 workers.
    - Average time to close out reported hazards.
    - Percentage of hazard reports leading to control implementation.
3. **Behavioral Safety Engagement**
    - Number of safety observations submitted by contractor personnel.
    - Toolbox talk completion rate.
    - Behavioral audit results (e.g., PPE usage, line-of-fire behavior).
4. **Incident Response Readiness**
    - Percentage of workforce participating in emergency drills.
    - Time taken to escalate a reported safety concern.
    - Post-drill performance rating.

5. **Contractor-Led Initiatives**
    - Number of contractor-initiated safety campaigns.
    - Quality rating of submitted safety improvement ideas.

This last category is important. It reflects whether the contractor is simply complying—or truly owning—its part in creating a safer work environment.

Incentives tied to KPIs work best when they reward contribution, not just outcomes. For example:

- Offer recognition for identifying and addressing safety risks, even if they are minor.
- Celebrate contractors who show year-on-year improvement, even if they start with a less impressive baseline.
- Reward transparency—reporting a near miss should be seen as a sign of maturity, not failure.

If you focus your KPIs solely on incident-free days, what you incentivize is silence. If you focus instead on hazard recognition, safety leadership, and learning behaviors, what you incentivize is vigilance.

YOU GET WHAT YOU MEASURE. AND WHAT YOU REWARD, YOU GET MORE OF

That principle isn't just a catchy phrase—it's the operating system of your contractor management strategy.

But here's the nuance we often miss: you only get what you measure *accurately*. And accurate measurement depends on systems, processes, and most critically—good data. Without it, your KPIs are not just ineffective—they're misleading.

Peter Drucker, the godfather of modern management, put it simply: "If you can't measure it, you can't improve it." But let's adapt that for our context: if you can't trust what you're measuring, you can't control it—and if you can't control it, then safety, compliance, and accountability become moving targets.

That's why leading indicators, for all their advantages, come with a necessary caveat: they are only as good as the data that feeds them. And that's where many organizations falter—not because they lack intent, but because they lack infrastructure.

Let's step back.

In the previous sections, we made the case for leading indicators: they are proactive, culture-shaping, and closely aligned with behaviors that prevent incidents before they occur. But unlike lagging indicators, which are rooted in hard outcomes (a fall, a lost-time injury, a near miss), leading indicators often track human activity: observations made, toolbox talks conducted, pre-job safety assessments completed.

That's all measurable—if you have the right systems in place. And more importantly, if those systems are used consistently, without manipulation, delay, or data gaps.

Consider this common scenario: your contractors are expected to submit daily safety observations. You use this data to calculate compliance and engagement levels. But the system used is manual—clipboards, paper forms, and occasional Excel uploads. How confident are you that those observations are being reported in real-time, with fidelity? What's stopping a site supervisor from backfilling the form at the end of the week, just to tick the box? And how do you know whether the observation had any meaningful follow-up?

Now imagine that same process digitized. Each observation is submitted through a mobile app, time-stamped, geo-tagged, and linked to a specific subcontractor and work activity. Photos are attached. Observations are reviewed by the site safety lead, who assigns an action. That action is tracked, closed, and analyzed.

Suddenly, your data becomes actionable. Your leading indicators transform from theoretical best guesses into real-time insight. That's the power of digitalization—not as a buzzword, but as an enabler of integrity.

Again! Let's look at how this plays out in the real world.

On one of the largest infrastructure projects in the world—the Riyadh Metro—Bechtel, as part of the BACS Consortium, implemented digital field tools to improve safety reporting and compliance. The project involved thousands of workers and multiple contractors across a complex urban site.

Using a digital safety management platform, the consortium tracked near-miss reports, job hazard analyses, safety conversations, and inspection findings in real time. This enabled centralized visibility into leading indicators and allowed senior managers to allocate resources where risk trends emerged. According to Bechtel's own reports, this approach significantly improved hazard identification and reduced the administrative burden on field supervisors.

Critically, it also reshaped the culture. Supervisors weren't just checking boxes—they were responding to real data, updated daily, with trends they could visualize and influence. And that is the essence of proactive safety leadership.

Chevron has taken digital safety a step further by integrating predictive analytics into its contractor management approach. Using historical and real-time data, the system identifies patterns that suggest an elevated risk—such as multiple contractors simultaneously working at height, or repeated permit-to-work deviations.

By feeding this intelligence to field managers, Chevron doesn't just monitor performance—they anticipate failures before they occur. Their system includes both leading and lagging indicators, but it's the real-time insights that shape frontline decisions.

Back to Your KPIs

What do all these examples have in common?

1. They don't rely on memory.
2. They don't reward reporting for its own sake.
3. They ensure that what gets measured gets managed—because it gets seen.

And this brings us to a crucial point: you can't design good KPIs in isolation. If your KPI is "95% of safety walks completed weekly," but you have no way to verify or timestamp those walks, it's meaningless. Worse—it invites gaming the system.

That's why smart organizations do this in phases:

1. **Establish the Need**: Define what behaviors or trends you want to measure.
2. **Build the Infrastructure**: Ensure you have digital tools, training, and support systems to collect accurate data.
3. **Test for Integrity**: Validate the quality and consistency of your data collection.
4. **Set the KPI**: Only once you're confident in the data flow and availability should you make that metric a performance requirement.
5. **Monitor and Adjust**: Leading indicators must evolve as project dynamics change. Static KPIs quickly become outdated or misaligned.

Let's be clear: digitalization isn't a magic wand. A poorly configured app or dashboard won't make your contractor workforce any safer. But when thoughtfully implemented, digital systems can:

- Standardize data collection
- Reduce reporting fatigue
- Enhance traceability and auditability
- Enable early intervention
- Support transparency across teams and suppliers

Most importantly, they allow you to treat your leading indicators as live indicators. You don't need to wait for a quarterly report to know something is off—you can act on a daily pattern, a flagged observation, or a slowing trend in near-miss reports.

Let's return to the broader theme of culture. KPIs aren't just targets—they are cultural cues. They tell people what matters. And if your indicators are real-time, behavior-focused, and transparently acted upon, then your contractors will respond not with resistance, but with engagement.

They'll understand that safety isn't something you only care about when an accident happens. It's something you're watching—proactively, visibly, and consistently.

Because what you measure, you value. And what you value, people protect.

By setting the right KPIs, you're not just measuring performance. You're shaping it.

HIGH RISK ACTIVITIES THROUGH PERMIT-TO-WORK SYSTEMS (PTW)

When it comes to high-risk work—such as hot work, confined space entry, working at height, or heavy lifting—the Permit-to-Work (PTW) system remains one of the most crucial tools in ensuring contractor safety. However, as with many systems, its effectiveness depends on how it is implemented. In too many cases, PTWs become little more than a formality—an exercise in paperwork that encourages shortcuts or complacency. When done right, however, a PTW becomes a living system of control, visibility, and accountability.

The purpose of a PTW system is simple: to clearly define the conditions under which specific high-risk work can proceed. It should outline the hazards, the control measures, the people responsible, the timing, and the location. Importantly, it creates a space for dialogue—a chance to pause, assess, and align before potentially hazardous tasks begin. But too often, that pause is rushed or skipped altogether.

In older systems—particularly those that are paper-based—permits can become bottlenecks. They're either filled out hastily, buried in site offices, or treated as box-ticking exercises. Contractors, under pressure to deliver quickly, sometimes see PTWs as an obstacle rather than a safety tool. And if the client isn't actively monitoring how the PTW process is applied, corners may be cut.

A modern approach to PTWs recognizes this. Digital systems, now widely available, streamline the process. They ensure that fields can't be skipped, that permits are logged and tracked, and that changes in site conditions can trigger permit reviews or revocations in real-time. These platforms also allow supervisors and safety officers to access and review active permits from their devices, offering visibility across multiple worksites.

But digitalization isn't just about convenience. It's about shifting the culture. When contractors know that their PTW submissions are being actively reviewed, when they're required to upload photos or answer specific hazard-related prompts, it encourages more thoughtful engagement. And when those same systems are used during walkdowns or toolbox talks, they become part of the workday rhythm—not just admin.

Consider the tragic example of Piper Alpha in 1988. While not a digital era case, it remains one of the most telling examples of permit failure. A pump had been removed for maintenance, and a blind flange meant to seal the system was not properly recorded on the permit. A night crew, unaware of the work, restarted the system, leading to a gas leak and subsequent explosion that killed 167 people. It was a communications failure. A permit failure. And it highlights what's at stake when PTWs are treated casually.

More recently, a contractor in New South Wales, Australia, was severely injured while conducting hot work on a fuel tank. No valid permit had been issued. Investigators found the risk assessment had been copied from a previous job and didn't reflect the specific hazards. The company was fined AU$180,000 and forced to overhaul its PTW process.

These are not just historical footnotes. They're reminders that high-risk work needs high-quality systems. A good PTW process ensures that:

- permits are issued and reviewed by competent personnel,
- are specific to the task and the conditions,
- are understood by everyone involved—not just signed, and
- are monitored while work is underway.

Permit systems aren't a substitute for supervision—but they are a reflection of it. They tell us whether risks are being considered, whether the contractor understands them, and whether the work should proceed at all. And in a modern contractor management framework, they serve as one of the most immediate interfaces between client control and contractor action.

In our experience, effective PTW systems are most successful when they're integrated with inductions, when they include digital sign-off, and when they're accompanied by real-time audits or spot checks. They shouldn't just be seen as a safety department task—they should be owned by site leadership, understood by frontline teams, and respected by contractors.

Because when they fail, they don't just slow a project—they can destroy it.

WHAT ELSE DO WE NEED TO MONITOR?

The last few years have seen a number of changes in monitoring projects from an HSE perspective. Nothing has been taken out, but some things are now monitored with great care that, in the past, really weren't looked at. They include welfare, well-being, human rights, ethical recruitment, and regular and accurate payment.

ETHICAL RECRUITMENT

It sounds nice, but ethical recruitment is not just an expression to be used to show that your heart is in the right place. Ethical recruitment is mandated by the International Labor Office which published International Labour Organization (ILO) Convention C181 to say what is and is not ethical recruitment when carried out by private recruitment agencies. The ILO should matter to all contractors because compliance with ILO mandates is increasingly a requirement either specifically for the project in question or more generally for the country in which the project is to take place. C181 also matters because, in construction projects of any size, almost all recruitment other than of the most senior management will be delegated to private recruitment agencies and not conducted directly by the contractor.

HUMAN RIGHTS OF MIGRANT WORKERS

Another important ILO document is the Declaration on Fundamental Principles and Rights at Work. Its fundamental points are:

- Workers must enjoy freedom of association and the effective recognition of the right to collective bargaining.
- There must be no forced or compulsory labor.
- There must be no child labor.
- There must be no discrimination in respect of employment and occupation.

There is also a requirement that people be paid on time and in the correct amount without deduction. On the face of it, this can at first sight seem like a problem in certain Middle Eastern countries and, in particular, in Saudi Arabia. Most projects of any size in the kingdom are ultimately funded by government, and government funds tend to go in peaks. The problem with peaks is that they are accompanied by troughs, and when money is short, the government stops paying. Saudi contractors are used to this and the chain of non-payment goes all the way down and embraces every employee. They simply don't receive the money they are due to receive. Any employee who finds this a serious problem can go to the company he works for and ask for an advance of part of his wages. Eventually, government money flows again, the bills are paid, and the shortfall in wages is made up.

Anyone who has worked for any length of time in the Middle East knows this is going to happen from time to time. Fine. But it simply doesn't work with migrant workers. A laborer on your site is almost certainly sending a chunk of his earnings back to Somalia, Ethiopia, Sri Lanka, Bangladesh, Nepal, or whichever country he calls home. That remittance makes a great difference in the lives of his family. They are able to feed themselves, to be sheltered and, increasingly, to give their children a better education than they had any hope of receiving themselves. But if the money dries up, they starve. And they are working for you. Which means that their family's starvation is something only you can prevent.

What this comes down to in the end is that migrant workers must be paid. In full and on time. And making sure that happens is your responsibility. Morally, ethically, and in every other way.

Your Team 7

Most of this book so far has been about the contractor—what the contractor does, who the contractor employs, how those people are managed. But what about the client? The owner? And the people who work for the owner?

Many of the same considerations apply. The natural instinct, because this is how it has always been done and these are the things that have always been important, is to follow the financial KPIs. Keep the overall cost within budget. Avoid delays. Make sure that quality targets are met and there won't be any disappointment when the project is complete and the site is handed back.

Of course, all of those things matter. But Health, Safety, and Environment (HSE) is now such an important part of project management that, if the HSE aspects of the contract are not looked after with the same care as everything else, the project will not in the end be rated a success. And, if you are the ultimate owner of the completed project, that failure will be yours.

So what do you have to do?

UNDERSTAND CULTURAL DYNAMICS

Culture is a force most safety professionals know is there, but too few dare to speak about. Not because they don't believe in its power. But because acknowledging that people from different parts of the world view risk, safety, and authority differently opens a difficult, and often politically charged conversation. It is easier and more comfortable, to pretend that culture can be flattened by a set of policies, that everyone will behave the same way if we write clear enough procedures and deliver enough toolbox talks. But reality says otherwise.

When we engage contractors from across borders, often thousands of them on megaprojects, it is not just their technical qualifications or work histories they bring. They bring values. Norms. Habits. Educational Environments.

DOI: 10.1201/9781003754121-7

Deep-seated ideas about hierarchy, fate, responsibility, and fear of shame. They bring culture. And if we ignore that, we aren't just missing an opportunity to strengthen safety, we are increasing our exposure to risk.

In 1980, Dutch social psychologist Geert Hofstede published his seminal work on national cultures in the workplace, outlining six dimensions: power distance, individualism versus collectivism, uncertainty avoidance, masculinity versus femininity, long-term orientation, and indulgence. These dimensions have since been applied extensively across disciplines, including occupational safety. Why? Because they shape how people follow rules, speak up, report hazards, or remain silent.

Cultural influence is not about right or wrong. It's about *fit*. A procedure developed in Sweden or the Netherlands, where low power distance and high individual autonomy dominate, may struggle in a workforce of contractors from South Asia, where hierarchy and collective obedience are cultural norms. If your safety system doesn't account for that, it won't matter how good the rules are. They won't land.

Japan has one of the lowest workplace fatality rates among industrialized countries. This success is often credited to a culture of meticulousness, team-based responsibility, and an almost obsessive commitment to order. However, it also comes with challenges, particularly when foreign contractors are introduced. In 2015, a Japanese highway project involving Southeast Asian contractors saw a rise in near-miss events. A post-project review found that non-Japanese workers felt hesitant to challenge unsafe actions taken by Japanese supervisors, citing fear of shame and disrupting group harmony (known as "wa").

South Africa's mining sector employs tens of thousands of contractors from diverse backgrounds. Cultural tension, shaped by decades of racial segregation and economic disparity, still echoes in many safety systems. Studies published in the *Journal of Occupational Health* reveal that Black South African miners are often reluctant to speak up about safety due to historical mistrust of management. Moreover, a strong masculine culture in mining discourages reporting injuries or expressing fear, which is seen as weakness.

The UK's Crossrail project we discussed in Chapter 4 had over 10,000 workers from more than 30 countries, managing cultural diversity was a critical safety challenge. A 2017 report by the UK HSE found that language barriers and cultural differences in communication styles led to a 25% higher rate of misinterpreted safety instructions among non-native English-speaking contractors. Workers from hierarchical cultures were also less likely to participate in behavioral safety observations.

In multiple studies across Latin American construction projects, researchers found that a cultural belief in "fatalismo" the idea that death and injury are predetermined, led to poor safety engagement. Workers were less likely to intervene in unsafe acts or report hazards, believing that outcomes were largely beyond their control. In one Peruvian mining project, this worldview contributed to several unreported near-misses that culminated in a major underground incident.

Here are some observations on how culture affects common health and safety practices:

1. **Toolbox Talks**: In collectivist cultures, public praise during toolbox talks may be more effective than individual recognition. In contrast, individualist cultures respond better to personal performance tracking.
2. **Stop-Work Authority**: In Sweden and Norway, nearly all employees report feeling comfortable using stop-work authority. In Indonesia and Pakistan, studies show less than 25% feel they can stop a job without fear of retaliation.
3. **Hazard Reporting**: A study by the International Journal of Environmental Research and Public Health found that Filipino workers in the Middle East were 60% less likely to report hazards than their Western peers, citing fear of deportation and cultural deference to authority.
4. **Safety Posters**: In cultures with low literacy or different symbolism (e.g., color meanings), standard safety posters may be misinterpreted or ignored. A Saudi Arabian contractor found success using pictograms instead of text, after injury rates failed to drop despite multiple awareness campaigns.
5. **Personal Protective Equipment (PPE)**: In parts of sub-Saharan Africa, wearing PPE such as masks or goggles can be stigmatized as weakness or distrust in one's skill. Portraying PPE as a sign of professionalism, helped change this narrative in Ghanaian gold mines.
6. **Leadership Engagement**: In high power distance cultures (e.g., China, India), the involvement of senior managers in safety walkdowns significantly boosts engagement. Workers are more likely to follow safety protocols if they see authority figures model the behavior.

The answer isn't to ignore culture. Nor is it to override it. It's to *engage* with it, intentionally, respectfully, and strategically.

That means:

- **Cultural Mapping**: Before projects begin, map out the cultural backgrounds of your contractor workforce. Use tools like Hofstede's Insights or consult with intercultural trainers.

- **Customized Induction**: Tailor inductions based on national cultural profiles. Use translators, visual materials, and culturally resonant storytelling.
- **Cultural Safety Liaisons**: Appoint team members who understand both the host and contractor cultures to mediate misunderstandings and build trust. This is significant particularly in Supervisory positions.
- **Feedback Loops**: Create anonymous channels for safety concerns, particularly in cultures where speaking up is discouraged.
- **Focus on Front-Line Supervisors**: Help front-line leaders recognize cultural cues and understand when silence isn't agreement … it's fear.
- **Celebrate Cross-Cultural Champions**: Recognize workers who bridge the cultural gap. Share stories of success and make them part of your engagement strategy.

Culture is not an excuse—it is a context. One that, if ignored, can derail even the best-designed safety program. But if understood and embraced, it becomes a multiplier. A catalyst for engagement, ownership, and shared responsibility.

We need to stop kidding ourselves that a one-size-fits-all approach to safety is enough. It's not. In fact, it's often part of the problem. The most effective organizations and the most capable safety professionals recognize that cultural intelligence isn't a nice-to-have. It's mission-critical. It's as essential to risk mitigation as engineering controls or PPE. Time and again, we've seen safety programs falter not because they lacked technical rigor, but because they failed to resonate with the people they were designed to protect. In contractor environments, this risk multiplies. Contractors bring with them a rich, complex mix of cultural norms, communication styles, and perceptions of authority and risk. Treating them all the same under a single model isn't just ineffective, it's negligent. If we want safety systems to stick, they need to be built with cultural fluency baked in from the start.

Understanding these dynamics isn't about stereotyping. It's about tailoring. Just as engineers account for environmental conditions when designing a structure, safety professionals must account for cultural conditions when designing systems meant to protect human lives.

When Chevron launched its operations at the Tengiz oil field in Kazakhstan, it encountered a cultural mismatch. Western-trained safety managers promoted open-door reporting, where workers were encouraged to stop unsafe work and challenge authority when necessary. However, Kazakh culture traditionally values hierarchical respect and deference to authority. Workers felt uncomfortable pointing out mistakes by supervisors or declining to follow potentially unsafe instructions.

Chevron responded by adjusting its training and communication programs. Rather than pushing a Western-style empowerment model, it embedded safety interventions into structured team meetings where concerns could be raised collectively rather than individually. Supervisors were coached to explicitly invite feedback and share stories about how lower-ranking employees had prevented incidents. By shifting the context, the company improved safety participation without forcing a cultural confrontation.

Toyota's global reputation for quality and safety owes much to its foundational principle of kaizen (continuous improvement). But the way this is practiced in its Japanese plants differs from its American counterparts.

In Japan, kaizen is deeply communal. Teams take responsibility together, and workers are culturally inclined to think in terms of group harmony (wa) we spoke about earlier. Safety concerns are raised in ways that maintain group cohesion. By contrast, in the US, a more individualistic approach prevails. American workers are more likely to raise concerns directly and seek individual recognition for improvements.

Toyota didn't insist on a uniform model. Instead, it maintained core safety principles while allowing local leadership to adapt processes to the prevailing culture. This flexibility helped preserve the integrity of its safety systems while making them locally effective.

When Shell developed contractor safety programs in the Niger Delta, they found that Western-style individual accountability frameworks did not translate well. Local communities valued communal decision-making, and risk tolerance was shaped by daily exposure to socioeconomic instability.

To succeed, Shell worked with community leaders and employed local safety champions who were respected within their groups. Safety training sessions became community events rather than corporate lectures. Accountability systems were redesigned to reward collective outcomes, such as team-based incident-free days, rather than just individual performance. This communal framing allowed Shell to integrate safety without alienating cultural norms.

The UAE's construction boom has relied heavily on migrant labor from South Asia, Southeast Asia, and Africa. These workers bring with them a wide array of beliefs about authority, safety, and responsibility.

One major construction firm found that workers from cultures with high power-distance indices (such as India, Nepal, Bangladesh, or Pakistan) were less likely to intervene or question unsafe instructions. Meanwhile, workers from low power-distance cultures (like the Philippines) were more comfortable doing so.

To accommodate this, the firm implemented peer-led safety briefings in multiple languages, led by trained safety champions from within each cultural group. They also modified their incident reporting system to allow anonymous reports, reducing the social risks of whistleblowing. The result was a measurable increase in near-miss reporting and early intervention.

Siemens encountered difficulties with its European-designed safety programs when managing wind farm projects in Latin America. Despite clear training modules and enforcement policies, uptake was slow.

On further investigation, Siemens found that language and trust were the primary barriers. Workers understood the technical material, but the programs lacked emotional resonance. Spanish-language translations often felt stiff or bureaucratic. Moreover, workers did not trust remote safety managers flown in from abroad.

Siemens revamped its approach. It hired local safety managers with deep cultural understanding and strong interpersonal networks. Training materials were rewritten not just in Spanish, but in region-specific dialects with locally meaningful examples. Roleplay and storytelling replaced PowerPoint slides. In follow-up evaluations, worker engagement with safety protocols increased by over 40%.

These examples offer more than anecdotes—they illustrate core design principles:

- **Language matters**, not just for clarity but for emotional connection.
- **Hierarchy and authority must be understood**, respected, or challenged appropriately.
- **Risk perception is cultural**, shaped by lived experience and social norms.
- **Group vs. individual orientation changes how accountability is internalized.**
- **Trust-building is not a side project**; it's the foundation of safety engagement.

Effective contractor safety programs don't erase these differences. They build frameworks that hold firm on principles (zero harm, stop-work authority, hazard identification) while flexing in their delivery.

To build systems that work across borders, companies must take the time to understand what motivates, what inhibits, and what resonates. That's not about being politically correct. It's about being operationally successful.

RECRUITING THE RIGHT TEAM

If you're never planning a project outside your own country, recruiting the management team you need will be fairly straightforward. But once you step across your own borders, life becomes rather different. You need a team that understands your own culture, both national and corporate, but also one that either is or can become familiar with the culture in the country where the project is to be. As anyone who has ever done it will tell you, expatriate work and the expatriate life are not for everyone.

Strangely, languages are not the most important skill your team needs. Yes, if they're going to be working in an Arabic speaking country then the ability to speak Arabic will be a help. And the same goes for speaking Greek in Greece, Portuguese in Brazil and, in Indonesia, the form of Malay generally known as Indonesian. But the real global lingua franca is English. 1.35 billion people worldwide speak English, and those who aren't native English speakers tend to be in the higher paid jobs and are therefore the ones your managers are most likely to deal with day-to-day. (The number who speak Chinese is greater, not least because of the huge difference between the population of China and that of everywhere else, but outside China and a limited number of other countries, your managers are likely to find the ability to speak Chinese of little value).

What, then, does an expatriate need? I once asked an Englishman who spent his working life in Africa, the Middle East, and Asia, "What do you need to do your job?" His answer? "A cast-iron stomach." And there is something in that. When they go abroad in your interest, your managers are likely to find not only that the food they can find to eat differs from what they are used to but also that meal times fall at different times of the day. So, yes, a willingness to eat what's available when it's on the table is fairly essential. But other things are more important.

Most important of all is an open, outward-looking attitude that does not believe that any nation has a monopoly of good ideas or knows the best way to do things. They must not have any tendency to say anything resembling, "That's not how we do it at home."

A road-marking contractor operating in Iraq had to work with water-based paint. Iraq had never used road-marking paint in which the solvent was water; they had previously only used chemical solvents. The paint needed to be thinned, and you thin water-based paint with, amazingly enough, water. No problem—Iraq isn't short of water. (Actually, it is, but desalinated ocean water is available and that was suitable for the purpose). But it couldn't be used, because the local Ministry of Transportation who had let the road-marking contract had a set of specifications that said that whatever thinner was used had to be supplied by the company that made the paint. The contractor's Middle Eastern representative (who was an English expatriate) asked for that to be waived in this case; the request was refused.

This paint had been manufactured by a company in Texas, USA. When they received the order from their representative, it specified the number of tons of paint to be provided *and the number of gallons of water.* They had never shipped water to another country. Why would they? They rang the representative and asked for an explanation—which they got. And they were told that the water should be shipped as it was—nothing added. But they weren't prepared to accept what the representative said. He wasn't American, and therefore his word didn't carry much weight. He may know Iraq, but he'd never been to Texas. In any case, this request came from Iraqis, and hadn't George Bush had trouble with those guys? In any event, what they actually shipped was water with something added to preserve the water.

No, I have no idea what it was preserving it against—I got this story from the local representative and he didn't know, so he couldn't tell me. But the additive was a chemical compound that a chemist would describe as a form of alcohol. You wouldn't drink it, but the customs officers at Umm Qasr, the port through which the paint and thinner were being imported into Iraq, read what was on the invoices and bills of lading, looked it up in whatever reference document they had available, and found that it was indeed a kind of alcohol.

Although Iraq is a Muslim country, it does not have the total ban on alcohol you find in, say, Kuwait or Saudi Arabia. Nevertheless, alcohol is controlled, and subject to religious disapproval. And what these customs officers had found was that an American company was attempting to bring alcohol into Iraq without a license and without declaring what it really was. Mayhem erupted.

I've chosen this story from many because it's such a silly example of how failure to understand another culture, and to accept that people who don't live next door to you may nevertheless know what they're talking about, can cause trouble.

And don't imagine that I'm talking only about Americans. In a Belt and Road Initiative infrastructure contract in Pakistan, a road was being built using Chinese funds and by Chinese engineers, but the work (and this is unusual for Chinese contracts) was being carried out by a local, family-owned, Pakistani company. The project did not go well. Management of the local company thought the engineers were treating them with condescension. And they objected. It didn't matter whether the Chinese were right or wrong—the Pakistanis found them patronizing and did everything they could to obstruct the engineers' wishes and not to do what the engineers wanted them to do. It took intervention at the highest level by governments in both China and

Pakistan to settle things down and get the road built. It would be very easy to blame the owner-managers of the Pakistani contractor—they adopted a very dog in the manger attitude and obstructed the Chinese engineers even when they knew that the engineers were right. But, actually, the Chinese did patronize the local guys, even though they may not have known that they were doing so. China has a particular way of doing things. Chinese culture is hierarchical. It says you defer to the most senior person, and you don't argue, even if you believe the most senior person may be wrong. Whatever changes in government and political philosophy the country may have seen, that has been true in China for at least 3000 years. (When China says it seeks harmony above everything else, it means what it says—just as long as you understand that this is the Confucian ideal of harmony, and that this kind of harmony only exists when people do what they are told). The engineers who had been sent from China to Pakistan were fairly new at expatriate life. They didn't yet understand and accept that other people in other countries did things differently.

The kind of people you want in your senior team are people who do understand and accept those things. You supplement them with local people. This can be very important. One thing that can help sell a contractor's bid is the promise to train local people, thus upgrading the skill levels of the local population. In fact, something that causes resentment against Chinese contractors is their tendency, wherever the job is that they are doing, to take all the staff with them from China and not to train any locals. Unfortunate—but it does help avoid the kind of problem we just described in Pakistan. If there are not sufficient local people who either have the necessary skills or are prepared to work on the project, you will have to use migrant workers.

In either case, training has to be done and some of that training may well need to be elsewhere. Perhaps in your own country. I'm not suggesting you move the whole migrant workforce to your own country for training before they start work—pick the brightest and most promising and train them so that they can carry out training when they get back. If you do that, you will have increased your costs but you will also have won the loyalty of the workers you chose.

SETTING STANDARDS

It's very easy for a chapter like this to descend into a series of exhortations. "Do this, and do it this way." "Don't do that." "Make sure everyone feels valued." And people would read that, and say, "Yes, yes, of course," and carry on doing what they usually do. Because that's how people respond to exhortation, and being told how to treat people. They regard these as apple pie issues. Things that are important—of course they are—but, look, we are building an airport here and the important thing is to make sure that the first flights can land and take off when they are supposed to. Everything else comes second.

But let's get serious about this. That airport you're building—isn't it more likely to come in on schedule if the workforce is on your side, shares your goals and feels pride in what it's doing? Rather than being indifferent to the end result? Or even being actively hostile because of the way they feel they've been treated? (And note that—what matters is not how they've been treated, or how you feel they've been treated—it's how they feel they've been treated. And those can be three different things.)

You want the whole workforce to buy in to your aims and values. To get that, you have to start with buy-in from the management team. They have to understand the need to build a culture in which every member of the workforce is valued and in which the vision is not simply to build the airport, or the road, railroad, building, sewage works, water treatment plant or whatever you happen to be engaged in. The vision takes in the idea of development. The development of the nation in which you're operating. The local development of the specific region. Development in which the workforce is going to become more skilled. And in which HSE will not only reach today's global state-of-the-art level but will also continuously develop as global standards develop.

Can you think of anything more desirable than to look back at the end of a working life and know you made the world a safer place for vulnerable people working in a hazardous environment? I can't. Make sure managers stay visible on site to engage with, praise, and monitor the workforce.

But that's only the beginning. The composition, behavior, and values of the client's team directly affect how contractors perform on the ground. In any large project, the client sets the tone. Whether by presence or absence, by clarity or confusion, by commitment or complacency—contractor behavior is shaped, encouraged, and in many cases mirrored by what the client's team chooses to prioritize.

Numerous studies—including work by the Institution of Occupational Safety and Health (IOSH)—have demonstrated that leadership visibility is one of the strongest predictors of positive safety performance. When client managers walk the site, engage meaningfully, ask questions, listen, and follow up—contractors notice. Conversely, when safety is only mentioned in project review meetings or reports but rarely witnessed on the ground, contractors quickly sense that production is the true priority.

A 2022 study published in the *Journal of Construction Engineering and Management* concluded that consistent client presence on site improved compliance behaviors by up to 30%. The researchers tracked behavioral observations across multiple infrastructure projects and found that client-initiated safety conversations—those started by someone from the client's team—had a cascade effect on contractor supervision.

Let's be clear: in many cases, the contractor's workers are not cutting corners out of malice or laziness. They're doing it because they think it's what the client wants. Rushed schedules, contradictory instructions, or the absence of enforcement suggest that the client is prepared to look the other way—as long as the job gets done. That's why accountability starts at the top. When a client project manager refuses to walk past unsafe conditions—or praises a subcontractor for stopping work to raise a concern—those actions ripple down the supply chain.

One of the more positive examples is the Heathrow Terminal 5 (T5) project. BAA, the client, implemented a framework called the T5 Agreement, which embedded behavioral safety expectations into contracts from day one. All project partners, including contractors and designers, were treated as part of one integrated team. Safety expectations were not bolted on—they were foundational. A key success factor? BAA's senior managers spent a significant amount of time in the field, reinforcing those expectations personally.

As a result, despite the scale and complexity of the project, T5 achieved industry-leading safety statistics for the time, with over 100 million work hours completed without a fatality.

Ironically, many clients invest in contractor training but forget to upskill their own teams. Project directors, procurement staff, planners, and site managers may not receive targeted training in how to support safe contractor performance.

One case study from a mining project in Chile showed that when the client implemented an HSE leadership course specifically for its own team—focusing on engagement, active listening, and just culture—the Lost Time Injury Frequency Rate (LTIFR) halved over 12 months. Why? Because their expectations became clearer, their behaviors more consistent, and their engagement more authentic.

Contractors often work in an environment defined by multiple layers of culture—national, organizational, team-based. The client's team needs to be culturally intelligent. What motivates the local workforce? How is respect shown? What are the unspoken rules of communication?

In West Africa, a development project supported by the African Development Bank struggled to enforce safety rules until the client team made a deliberate effort to understand and engage with the regional work culture. They translated safety communications into local dialects, appointed community liaisons, and even revised some procedural language to make it resonate with local values. The turnaround in workforce engagement and hazard reporting was dramatic.

It's hard to overstate the role of trust. A client team that creates a punitive or overly hierarchical environment will receive filtered or manipulated information. Safety risks get buried. But when the client fosters trust—by acting with fairness, transparency, and integrity—contractors are far more likely to surface concerns early.

In 2019, a Singaporean public housing development created a system where contractors could anonymously report client-induced safety pressures, such as unrealistic timelines or scope changes without proper safety review. This feedback loop allowed the project leadership to correct internal issues that were inadvertently driving unsafe behavior. It wasn't just about monitoring contractors—it was about reflecting on their own role in the system.

Ultimately, the client's team is the most powerful cultural force on a project. More than policies or penalties, it is the client's example that sets the boundaries of what is acceptable. A client that listens, that values people, that acts consistently and with vision, will see that echoed in contractor behavior.

This is not sentimentality. It's strategy. Because no contractor management system, no induction video or permit form, will ever be as powerful as the lived experience of seeing a client leader care about the same things they ask you to care about.

So yes—walk the site. Talk to the people. Listen more than you speak. Clarify your expectations. And build the kind of culture that doesn't just finish the project, but leaves behind a legacy of safety, respect, and pride.

Because if you can shape how contractors feel, you can shape how they work. And when that alignment happens—when values are shared and goals are common—that's when great projects deliver safely, proudly, and on time.

INVOLVE GOVERNMENT

You and the people who work for you are going to be in this place until the project is completed and you've handed it over. Then you will leave. And the people who live here, or who move in as a result of your project, will still be here. You have ownership of HSE matters as long as you're here. You are responsible for them. But, after you've left, that ownership and that responsibility pass to local government bodies. So get them involved from the start. Make sure they understand the HSE initiatives you implement, and why you implement them. And while you're about it, make sure the local press know too.

Your team may be made up of some of the best professionals in the field. But when they depart, so does their institutional memory. Government bodies will remain, and in many ways, they are your project's true successors. If they are not engaged, trained, or informed, the legacy of your hard work can vanish the moment the ribbon is cut.

It's easy to fall into the trap of managing everything yourself because it's quicker, cleaner, and often more efficient in the short term. But if your safety systems rely entirely on your team's presence, then you haven't built sustainability—you've built dependency.

Effective contractor management in partnership with local authorities doesn't just safeguard the workforce today—it prepares those responsible tomorrow to understand what's been done and why it matters.

A compelling example of successful government-contractor collaboration in safety management is the US Department of Energy's (DOE) approach to managing contractor safety on hazardous waste cleanup projects. A study published in the *American Journal of Public Health* analyzed the DOE's safety culture and its impact on contractor safety performance.

The study found that while the DOE had an excellent safety culture on paper, inconsistencies in procurement practices, and contract language led to mixed messages regarding safety priorities. These inconsistencies resulted in underreporting of incidents and underachievement in safety performance among some subcontractors.

The takeaway? If government agencies want safety to be real, not just rhetorical, they must align their procurement policies and contract oversight with their safety ambitions. This means that language around safety in tender documents, monitoring requirements, and enforcement mechanisms must all reinforce the idea that safety isn't optional—it's foundational.

Another illustrative case is Tata Steel's initiative to enhance contractor safety management. In collaboration with dss+ Consulting, Tata Steel implemented a comprehensive contractor safety management framework that emphasized leadership commitment, consistent safety practices, and inclusive engagement of both employees and contractors.

What's less often discussed, but equally critical, is how this initiative was aligned with Indian regulatory bodies and regional safety authorities. By ensuring that government safety inspectors and officials were familiar with their systems, Tata Steel didn't just improve their own performance—they helped raise the bar across the region.

In Brazil, the government's Growth Acceleration Program (PAC) has faced scrutiny for safety lapses on major infrastructure works. A 2017 audit found that while contractors were required to meet certain national HSE standards, few government bodies were adequately trained to enforce them. The result? A disconnect between what contractors committed to on paper and what happened on site.

In response, Brazil's Ministry of Planning partnered with international agencies to train inspectors, align HSE performance indicators across sectors, and even introduce digital monitoring tools shared between contractors and regulators. This effort not only improved compliance but strengthened trust between public bodies and the private sector.

In Sarawak, Malaysia, the Sarawak Corridor of Renewable Energy (SCORE) project included multiple high-risk construction and energy ventures. One of the lessons learned early on was that national safety standards were not consistently understood at the local district level. Contractors from outside the region found themselves dealing with inspectors unfamiliar with permit-to-work (PTW) systems or modern hazard identification protocols.

The solution? Joint workshops involving contractors, client HSE leads, and local government safety officials. These sessions focused not just on policies but on lived examples—photos of scaffolding errors, near-misses during excavation, and how better collaboration might have prevented them. The impact was immediate: clearer expectations, better site audits, and smoother project approvals.

After the London 2012 Olympics, much was made of how safe the construction phase had been—over 46 million work hours with no fatalities. What is less talked about is how the safety legacy was handed off to local boroughs and agencies post-games.

A similar, though contextually distinct, transition unfolded in Qatar following the FIFA World Cup 2022. The lead-up to the tournament involved a massive infrastructure expansion—stadiums, transport corridors, accommodation zones, and energy upgrades—all constructed under intense international scrutiny. While much of the media attention focused on worker welfare during the construction phase, what is less widely discussed is how the Supreme Committee for Delivery and Legacy (SC) prepared for the post-event safety legacy.

In the years preceding the World Cup, the SC developed a tiered contractor HSE monitoring framework, working with international consultants and certification bodies. This system included monthly safety performance audits, incident classification reviews, and mandatory leadership safety walkabouts. What made it significant was the deliberate capacity-building element—Qatari government inspectors from entities such as the Ministry of Municipality and the Ministry of Labour were seconded into SC project teams. This allowed knowledge transfer to occur in real-time, not in theory.

As the tournament concluded, many of these inspectors returned to their ministries—but now with a deeper understanding of global HSE standards, practical application of behavior-based safety, and modern contractor engagement models. In essence, the post-World Cup handover was not just operational—it was institutional. The safety documentation systems, the training platforms, and even the analytical tools used to manage contractor performance were transitioned to various governmental departments and retained for use in national infrastructure projects.

Qatar's Ashghal (Public Works Authority) and KAHRAMAA (Qatar General Electricity & Water Corporation) are examples of state agencies that benefitted from this transition. Lessons from stadium safety programs—such as real-time PTW tracking, multi-language induction systems, and digital fatigue monitoring—have since been adapted for use in ongoing metro, water, and smart city projects across the country.

It's a living case of how mega-events can leave behind more than just physical infrastructure. They can reshape regulatory mindsets, equip governments with better systems, and create a more resilient contractor safety environment—provided there is intentional collaboration between client teams and local authorities from the beginning.

The Olympic Delivery Authority in the United Kingdom as well as the Supreme Committee for Delivery and Legacy (SC) in Qatar, working in tandem with local authorities, created a transitional HSE handover document. This included:

- a summary of key risks,
- safety-critical infrastructure,
- community safety engagement strategies, and
- a registry of lessons learned and recommendations for future developments.

This sort of structured government engagement ensured that the culture built during the high-performance phase didn't vanish once the stadiums and parks opened to the public.

In many regions—especially where international projects are implemented—there is a risk that HSE language becomes technical, imported, and inaccessible. If government officials don't understand the terms being used (e.g., "LOTO," "Permit to Work," "dynamic risk assessment"), they cannot play an effective role in oversight.

The Baku-Tbilisi-Ceyhan pipeline project dealt with this by creating multilingual, pictorial HSE manuals designed for both local workers and government inspectors. This not only helped raise the capacity of contractors but provided a lasting resource for local government safety programs after project completion.

When government stakeholders are brought in early—not just informed, but involved—they often become champions of your work. This public endorsement has ripple effects:

- Contractors feel validated and perform better.
- Communities trust the project more.
- Media coverage is more constructive than critical.

Going back to the Crossrail project in London, their safety innovation—like vehicle proximity sensors and fatigue tracking tools—were co-developed with UK transport agencies. The result? These innovations didn't die with the project. They became part of ongoing UK transport policy.

When the ribbon is cut and you move on to the next project, what stays behind isn't just concrete or steel. It's the relationships, systems, and culture you helped build. Engaging local government is the most sustainable way to protect and extend that legacy.

Whether through joint training, co-developed procedures, real-time data sharing, or public communication campaigns—making local agencies part of your team creates resilience. It ensures that what you build doesn't just last—it evolves.

Five questions to ask today:

1. Have local government HSE officials been invited to planning meetings?
2. Do your HSE communications and reporting structures include government stakeholders?
3. Are public agencies familiar with your safety systems (e.g., PTW, incident tracking, risk assessments)?
4. Have you identified capacity-building opportunities for local inspectors or policymakers?
5. What will happen to your safety systems when your team leaves?

If you can't answer these questions clearly, your project may still be vulnerable to breakdowns post-handover.

Involving government isn't just best practice—it's future-proofing. It's responsible, it's ethical, and in many parts of the world, it's increasingly becoming a legal expectation.

Because long after the contractors go home, the road, hospital, power plant, or school will still be here. And the question won't just be "Was it built safely?"—it will be, "Is it still being kept safe today?"

Striking a Balance in International Contracts

8

International contracts can be confusing, because the people working on them tend to be people who choose to work on international contracts and not the workforce of a particular company involved in the contract. That means you can get a whole range of nationalities, each combining the culture and mindset they grew up within their home countries with the culture and mindset they have developed working in other people's cultures. It's a mind-opening experience, whatever else it may be.

DON'T LEAVE EVERYTHING TO THE CONTRACTOR YOU HIRED

"Everything" in this context includes liaison with local regulators. Things go wrong. When that happens, the most important thing is to put them right—but, for your contractor, putting things right may come second to preventing you from finding out what went wrong. That's why it's a good idea to maintain contact with the regulators—not out of some sneaky wish to check on the people you are paying to do a job for you, but to make sure (as far as possible) that you are not going to find yourselves paying even more to cover the cost of a contractor's errors or neglect.

And then there are simpler things in which you can be involved. A European company described the work they'd hired a local contractor to carry out for them in the Kingdom of Saudi Arabia (KSA). As so often in KSA,

DOI: 10.1201/9781003754121-8

the contractor's owners were local, the managers were a mix of Jordanians and Palestinians (and Palestinians traveling on Jordanian passports), and the laborers were from a number of African countries with Somalia and Sudan prominent among them. They did an excellent job. The European company had an individual on site to supervise. When the work was finished and he saw how good it was, he went round the crew shaking each man by the hand and saying, "Well done. Thank you." He could see one of the managers watching him and it was clear that the manager was thinking, "Why is he doing that? These men are of no importance." And if he hadn't been able to see that, he would have known anyway because the manager said almost precisely those words to him. "Why do you thank the crew? They don't matter. They do what we tell them to do."

But they do matter. According to the European observer, it was clear that the men were surprised to receive these thanks, but equally clear that they were pleased. And they didn't forget. Because, about a year later, the European company hired the same Saudi contractor for a new job and, when the crew turned up, they remembered the observer. He was greeted with nods, welcomes, and smiles. And the job they did for him this time was even better than it had been the first time.

People matter. They don't matter with the same intensity in all cultures. But the more they matter to you, and the more they know they matter to you, the better the job they will do for you.

Finally, there's the question of who is funding this project. Which brings us on to:

WHO IS PROVIDING THE FUNDING?

Not every project is paid for entirely (or, in some cases, at all) by the owner. Depending on where it is, who will benefit from completion of the project, and what the project involves, funding may come from a number of sources. They might include the World Bank, the International Monetary Fund, and a number of regional development bodies such as the European Regional Development Fund, the African Development Bank, and the Asian Development Fund. They all attach stipulations to their funding offers. Many of these are concerned with avoiding corruption and the misappropriation of public funds, and it is vital that your accounting is both accurate and timely—if you don't get your reports in on time, funding will be at best suspended and in some cases canceled. But others deal with the way the project is conducted and the standards applied.

WHAT STANDARDS APPLY?

As mentioned in the last paragraph, funding organizations will have a say in the standards to be observed by the project—but there are others. If your project is in infrastructure or building construction, there will almost certainly be a government ministry or department in the country where construction is taking place that sets the standards that have to be observed in that country. And the secret of that sentence is in its wording: they have to be observed. You have to abide by them. Even if they seem absurd. (You want absurd? The government of Cyprus, like most other countries, lays down standards for thermoplastic markings used on its roads. One standard it sets is that the thermoplastic must be capable of withstanding the impact of studded winter tires or tires with winter chains. The climate of Cyprus is classified as subtropical, Mediterranean, and semi-arid. Winters are very mild; summers range from warm to hot. Snow does sometimes fall in winter, but only high in the Troodos mountains. In other words, you will never see studded tires or tires with chains—but thermoplastic manufacturers must certify that the product they supply meets that requirement, even though doing so add significantly to the cost of production).

How does that come about? Like most countries, Cyprus may adapt standards to its own requirements but they obtain the major part of their national standards from international standards provided by International Organization for Standardization (ISO).

ISO (INTERNATIONAL ORGANIZATION FOR STANDARDIZATION)

You need to check that the country in which your project is taking place mandates ISO standards. You also need to remember that the national standards mentioned in the previous section take precedence—but this is usually simpler than it might be because the national standard will almost always be the relevant ISO standard with the addition, modification, or substitution of whatever clauses the national authority wants you to follow.

ISO standards are increasingly a help in winning business, because more and more companies are requiring confirmation of ISO compliance before accepting bids. Compliance isn't something a company decides for itself—you call on a specialist organization that will go through all your documentation, examine each one of your processes, and then watch what happens in practice to ensure that the written procedures are actually followed and that they reach the level required by the ISO standard.

ISO 45001 Health & Safety Management

As a recap to what we have already touched on, ISO 45001 is designed to help businesses that comply with it implement a system for managing health and safety and reduce the risk of accidents, claims, costs and business interruptions.

To become certified as ISO 45001 compliant, all of your systems and procedures will be assessed, and then the effectiveness with which they are implemented will also be assessed. If you've done everything described in this book, and if it is all working as intended and no-one is cutting corners, you should achieve certification without any difficulty. However, the real benefit of any ISO assessment, including this one, is often in the feedback you get from external inspectors who can either validate or question your belief that you are observing Health, Safety, and Environment (HSE) best practice.

ILO (INTERNATIONAL LABOUR ORGANIZATION)

The International Labour Organization's *Safety, Health and Welfare on Construction Sites* manual remains one of the most practical global guides for anyone managing contractors. Its message is simple: construction work is inherently hazardous, but most risks are predictable—and therefore preventable. When welfare, planning, and controls are treated as core operational requirements rather than optional extras.

Chemical hazards remain one of the most persistent threats on construction sites, appearing in paints, solvents, adhesives, fuels, and cleaning agents. The ILO stresses the hierarchy of control: eliminate or substitute dangerous substances wherever possible; engineer the hazard out through enclosure or ventilation; and only then rely on personal protective equipment (PPE) as the final barrier. Containers must be labeled, spills controlled immediately, and storage kept secure. The same disciplined approach applies to fire risks. If flammable liquids cannot be avoided, fresh air must be guaranteed.

Environmental exposures such as heat, cold, noise, or poor lighting quietly undermine safety long before an incident occurs. Protecting workers from these conditions is not comfort; it is competence. PPE remains essential, but the ILO reminds us that it should never compensate for unmanaged hazards. Hands, skin, eyes, feet, and the head remain the frontline casualties of poor planning.

Working at height requires harnesses and secure platforms, not improvisation. Respiratory protection must be provided wherever dust, mist, or gases are present.

Above all, welfare is not a luxury. Proper lavatories, washing stations, showers, drinking water, shaded meal areas, and hygienic food facilities are foundational to dignity and performance.

ESG AND THE FUTURE OF CONTRACTOR MANAGEMENT

Over the past two decades, ESG—Environmental, Social, and Governance—has shifted from a fringe concept into a core pillar of how organizations are evaluated, funded, and held accountable. Originally rooted in investment ethics and sustainability frameworks, ESG emerged as a practical tool for financial institutions to measure a company's long-term risks and value. But its impact today is far broader—reshaping everything from boardroom priorities to how procurement and contractor management are executed on the ground.

The origins of ESG can be traced back to a 2004 UN report titled "Who Cares Wins," which called on major investors to consider environmental and social factors in financial analysis. Since then, frameworks like the UN Principles for Responsible Investment (UNPRI), the Global Reporting Initiative (GRI), and more recently the EU Corporate Sustainability Reporting Directive (CSRD) have cemented ESG into mainstream operations. Today, global corporations and even SMEs are expected to demonstrate how their operations—not just their profits—create value for society, the planet, and their people.

So where does health and safety fit into this evolution? In the past, HSE was often a standalone function, focused on compliance, injury prevention, and risk management. But under ESG, safety performance is no longer just a legal or operational issue—it is now an ethical and reputational one. When companies report on their ESG performance, serious safety failures are seen as breaches of social license, failures in governance, and—in high-carbon or high-risk industries—failures in environmental responsibility.

As a result, contractor management is undergoing a major shift. Traditionally, clients assessed contractors based on cost, timelines, and technical capability. Then came safety prequalification. Now, ESG-conscious clients are asking:

- What is your contractor's carbon footprint?
- Do they align with human rights standards?
- Do they have inclusive employment practices?
- How do they monitor and report their own subcontractors?

And crucially: do they operate within an integrated, values-driven framework that reflects our ESG commitments?

This has direct implications for safety teams. It means that health and safety professionals can no longer work in silos. Permit-to-work (PTW) systems, risk assessments, and site audits must now connect to wider ESG dashboards and reporting tools. Contractors may need to demonstrate not just injury rates but how they contribute to net-zero goals, community engagement, fair labor practices, and diversity metrics.

In forward-looking organizations, the HSE department is evolving into a broader ESG role. We've seen safety leaders retitled as ESG Directors. Monitoring systems that once focused solely on incidents and near misses now track emissions, ethical sourcing, and labor standards. Training programs that once centered on PPE are now incorporating climate resilience, cultural sensitivity, and ethical leadership.

This isn't just theory. Consider the example of Vinci Construction in France. The company has integrated ESG into its contractor selection process, requiring subcontractors to meet sustainability and ethics criteria before bids are even reviewed. Their project dashboards include not only safety indicators but CO_2 tracking and workforce diversity ratios. This puts safety performance in the same arena as environmental and social performance—not separate, but interconnected.

Or look at the Qatar 2022 World Cup infrastructure program, where ESG scrutiny prompted a fundamental rethink in how contractors were monitored. The Supreme Committee for Delivery & Legacy implemented a Workers' Welfare Standards framework that blended traditional HSE monitoring with human rights audits, third-party labor inspections, and grievance mechanisms. This holistic model was arguably one of the earliest large-scale ESG applications in construction contractor management.

To stay relevant and resilient, contractor management systems must evolve. This includes:

- ESG Prequalification: Contractors should be assessed not just on technical or safety grounds, but on their ESG maturity. Do they publish sustainability reports? Do they meet local and international human rights standards?
- Integrated Reporting: Safety KPIs should feed into ESG dashboards that are visible to both internal stakeholders and external investors.
- Digital Compliance Tools: Platforms like EcoVadis, ISNetworld, and Avetta are already offering ESG ratings and compliance tools that include safety, labor, and environmental performance in a single score.
- Training and Culture: Contractors need onboarding that speaks to ESG culture, not just job-site hazards.

As ESG continues to shape investment flows, client expectations, and government regulations, contractor management must move beyond compliance. It must become anticipatory, ethical, transparent, and values-driven. For health and safety professionals, this is not a threat—it's an opportunity. By aligning with ESG, safety leaders can elevate their impact, influence strategy, and secure the long-term well-being of both people and projects.

Because in the ESG era, a safe workplace is not just a legal duty—it's a statement of who you are and what you stand for.

CYBERSECURITY AND DIGITAL RISK

As PTW systems and remote monitoring move deeper into digital platforms, an often-overlooked risk emerges: cyber vulnerabilities. Today, contractor safety processes aren't just paperwork—they're digital workflows managed via tablets, cloud-based dashboards, and connected sensors. While these advancements drive efficiency and visibility, they also open new avenues for threats that can directly impact physical safety.

A high-profile example occurred in the **SolarWinds cyberattack of 2020**. Attackers inserted malware into widely used network-management software, which then propagated across thousands of organizations—including government agencies and private corporations. Though the initial intrusion wasn't tied to safety systems, it exposed how third-party vendor access can cascade into governance and operational silos. Several organizations had their safety-critical infrastructure and data management compromised through vendor channels that were insufficiently segmented or monitored.

Out of reach? Not necessarily. Even safety-specific systems—like remote access to site CCTV, IIoT sensors, or mobile PTW apps—can be manipulated if contractor credentials aren't properly managed. A compromised device or identity can lead to:

- **System Sabotage**: Changes in entry permit status, false "work complete" signals, or disabled alarms.
- **Data Manipulation**: Erasing incident records or hiding near-miss reports.
- **Loss of Safety Context**: Removal or encryption of audit logs that inform operational assessments.

In 2019, the multinational aluminum producer **Norsk Hydro** was hit by a ransomware attack that impacted its European and US operations, including contractor-accessible systems. While their core plant safety systems remained insulated, the breach temporarily paralyzed digital permit tracking, third-party job scheduling, and asset management systems—causing work stoppages across contractor crews. The event highlighted how **cyber interruptions in support systems can directly undermine contractor mobilization and safe coordination**, even when physical safety equipment remains intact.

Modern safety systems must treat digital access as part of overall risk management. Here's what top organizations are doing:

- **Role-based, time-boxed access**

  Contractors only receive credentials relevant to their task window—for example, live access to their own permit until work completion. Afterward, the account is suspended.
- **Multi-factor authentication (MFA)**

  Especially important for any application connected to safety-critical systems.
- **Compartmentalized dashboards**

  Contractors should not access wide safety dashboards. Instead, they receive task-specific interfaces, limiting lateral movement.
- **Automated revocation**

  Temporary access expiration should be built into workflows—leveraging system triggers tied to job closure or inactivity.
- **Dual sign-off for high-risk permits**

  Even digital permits should require both contractor initiation and client sign-off, ideally from separate devices.

Contractor inductions, toolbox talks, and site briefings now need a digital dimension: best practices for device security, phishing vigilance, and password hygiene. Brief safety leaders should ask:

- Are contractors using personal devices to access safety systems?
- Are credentials shared loosely or stored insecurely?
- Is there visibility across devices logged in?

As safety and digital systems increasingly converge, your responsibility as a client is not just to ensure that contractors do PTWs correctly, but also that the systems enabling PTWs are themselves secure. Cyber hygiene, segmented access, real-time monitoring—and resilient backup plans—are now essential layers of contractor safety itself.

ARE THE STANDARDS PRESCRIPTIVE, OR GOAL-SETTING?

The difference between prescriptive and goal-setting standards is that prescriptive standards are laid down by relevant regulations and laws and you have to follow them; goal-setting standards describe current best practice. While it's important that you identify which is which, it would not be true to say that you only have to follow prescriptive standards—best practice should be everyone's aim.

Prescriptive standards are like the rulebook. They tell you what to do and how to do it. Think of them as minimum requirements—the legal floor. They specify, for example, the exact number of fire exits, the maximum load capacity of scaffolding, or the decibel limits for workplace noise. If you breach a prescriptive standard, you're in violation of the law. Simple.

Goal-setting standards, on the other hand, are more about outcomes than instructions. They ask: what are you trying to achieve? And are you doing everything reasonably practicable to achieve it? These standards are designed to encourage continuous improvement and innovation. They don't spell out every detail, but they do ask that you meet a benchmark of performance or behavior that's consistent with the current state of knowledge and good practice in your industry.

This is particularly important in contractor management because prescriptive regulations are often slow to adapt to changes in technology, materials, or methods of work. Goal-setting standards can fill this gap by encouraging organizations to go beyond compliance—to lead rather than follow.

In many jurisdictions, as an example, confined space work is governed by prescriptive rules: permit systems, air testing, rescue plans. But best practice goes further. In the UK, for example, the Health and Safety Executive recommends using the hierarchy of control to eliminate the need to enter confined spaces at all—something not required in law but seen as good practice. Similarly, Australia's WHS Act supports a goal-setting framework where employers must do "what is reasonably practicable" to ensure safety—even if that means exceeding legal requirements.

In modern contractor management, particularly on high-risk or fast-moving projects, the blend of both approaches is essential. Regulations provide the baseline—but your duty of care, your leadership, and your reputation are measured against best practice.

The organizations that thrive in contractor safety today are those that ask, not just "What must we do?" but "What more can we do?" It's in this space—between law and leadership—that the best safety cultures are built.

MAKE SURE EVERYONE KNOWS THE STANDARD YOU'RE WORKING TO

Really, this should be obvious—but we're stating it anyway: find the standards that apply to your project in the country in which you are operating and with the funding agencies who are putting up the money, and ensure that everyone involved in the project knows which standards you are working to and what they say.

This isn't just about bureaucracy. It's about alignment—between your team, your contractors, your regulators, your clients, and sometimes, your insurers. Misunderstandings about what standards apply—whether national law, funding body requirements, or internal company policy—can lead to confusion, inefficiency, and, in the worst-case scenario, serious safety failures.

In large infrastructure projects or multinational ventures, it's common for multiple standards to compete for attention. For example, a road being built in Kenya with funding from a European development bank may be subject to Kenyan national safety laws, the bank's environmental and social safeguard policies, and the contractor's own global standards. Which of these takes precedence? What happens if they conflict? And are all parties—even the subcontracted electrician or crane operator—aware of what's required?

The Bhopal disaster we discussed earlier remains a tragic reminder of what happens when standards aren't understood, enforced, or harmonized. An underlying cause to add to our previous discussion is that Union Carbide Corporation (UCC) had internal safety protocols in the United States that far exceeded what was practiced in the Bhopal plant in India. Despite being a UCC subsidiary, the Indian plant did not implement the same maintenance schedules, safety systems, or hazard controls.

Part of this came down to cost-cutting. But another issue was lack of clarity around which standard applied and who was responsible for ensuring compliance.

Many large-scale projects today—especially those backed by international financing—are subject to the **International Finance Corporation (IFC) Performance Standards**. These standards cover environmental and social sustainability, including Occupational Health and Safety. When a project receives IFC or World Bank Group funding, contractors must not only follow local laws, but also meet the higher expectations of the funders.

In a 2018 audit of a hydropower project in Southeast Asia funded by multiple international banks, it was found that while the local contractor followed national Occupational Health and Safety laws, they fell short of IFC standards in risk communication and community safety. The contractor was instructed to revise its HSE plans, re-train its workforce, and engage in direct dialogues with affected communities. The lesson? When financing comes with strings attached, your standards are no longer defined solely by geography—they become global.

In the Gulf region, a common challenge arises from the interplay of local labor law, free zone regulations, and client-imposed standards. For example, in Qatar or the UAE, a project in a government-owned economic zone might follow different HSE rules from one under a private developer—even though they are geographically side by side.

A multinational engineering firm working in the UAE on a metro project noted that confusion arose among subcontractors when the client required adherence to UK Construction Design and Management (CDM) regulations as a condition of contract, even though the project was not in the UK. Without sufficient onboarding, local subcontractors—unfamiliar with CDM concepts such as the Principal Designer role or the pre-construction information pack—delivered substandard safety planning. The main contractor responded by developing a hybrid training module that introduced international standards in a localized way, ensuring clarity and consistent application.

Having a well-defined and well-communicated standard does more than protect you legally. It also builds trust across your supply chain. When contractors understand the expectations upfront—preferably during the tendering stage—they can plan properly. This is especially critical in joint ventures or consortia, where different partner organizations may each be bringing their own standards to the table.

In South Africa, the renewable energy sector has adopted a clear matrix system to manage this complexity. Projects define a hierarchy of standards, where the highest (often client-imposed or lender-required) takes precedence. This matrix is shared with all contractors and is embedded into all project induction materials. One project developer, operating multiple wind farms, credits this transparency with reducing non-conformance reports by 45% in a single year.

Simply sending out a PDF of your HSE standard isn't enough. Effective communication means:

- **Training**: Dedicated sessions to walk teams through the standards that apply, including real-world examples.
- **Sign-Off**: Requiring contractors and supervisors to formally acknowledge their understanding of applicable standards.
- **Induction**: Embedding standard awareness into every onboarding and site entry process.
- **Ongoing Engagement**: Toolbox talks, visual aids, and field verification to reinforce what "compliance" looks like in practice.

In 2022, a construction project in Latin America funded by a multinational consortium encountered a shutdown after it was discovered that local contractors were not meeting the environmental mitigation standards required by one of the funders. The issue wasn't deliberate non-compliance—it was lack of awareness. The standard had been mentioned in the contract, but never clearly explained or integrated into the site management plan. As a result, downstream communities were affected, and reputational damage to the funding organizations ensued.

The remedy? The consortium revised their contractor engagement model to include a "standards onboarding" week before mobilization. This included live translation of technical standards into local languages and face-to-face discussions to clarify expectations.

When people say "we didn't know," it's rarely an excuse—it's a warning. Don't assume that the presence of a legal clause or a funding condition means it's been understood. Don't assume your contractors will read between the lines. And don't assume that standards are static—they evolve. What's considered best practice this year might be obsolete next.

If you want safe projects, start with clarity. Define the standards. Share them widely. Train against them. Embed them. Then check—constantly—that they're being followed.

Because when everyone knows the standard you're working to, it's not just compliance—it's competence in motion.

HARNESSING TECHNOLOGY TO ENHANCE HSE PERFORMANCE

In recent years, the fusion of digital innovation and safety culture has fundamentally reshaped how organizations manage HSE practices. Across industries, from oil and gas to manufacturing and construction, companies are leveraging wearables, artificial intelligence, virtual training environments, and real-time communication technologies to proactively identify risks, protect workers, and enhance operational efficiency. Let's explore how these tools are being applied in a way that is transforming contractor safety management.

WEARABLES FOR REAL-TIME RISK AWARENESS

At the forefront of innovation is the adoption of **HSE wearables**—devices that go far beyond fitness tracking. Take for example **Eni**, the Italian energy giant, which collaborated with researchers at the Massachusetts Institute of Technology to develop smart vests for workers. These vests don't just track vital signs like heart rate and temperature; they continuously scan environmental conditions such as gas presence and heat stress. This holistic monitoring creates a layered picture of health and risk, allowing real-time interventions.

Similarly, British Petroleum (**BP**) introduced wearable technologies for crews working in high-stress zones like the Arctic and Arabian Gulf. Here, environmental extremes can rapidly impact a worker's health. These vests, integrated with communication tools, not only monitor safety but ensure connectivity, reducing isolation in remote areas.

Marathon Petroleum has taken this further by equipping teams with wearable gas detectors that trigger instant alerts in case of toxic exposure. These wearables also track location, which becomes critical during evacuation or emergency rescue.

Such tools do more than measure; they act. And perhaps more importantly, they underscore a shift toward empowering workers with real-time situational awareness—helping them make safer decisions in the field.

AI IN ACCIDENT PREVENTION AND BEHAVIORAL ANALYSIS

Advances in artificial intelligence are giving safety managers a new set of eyes—ones that never blink. **TotalEnergies**, for instance, has implemented an AI-enhanced CCTV system that uses computer vision to monitor and analyze worker behaviors on offshore platforms. The system is trained to detect high-risk actions such as working at heights without harnesses, entering restricted zones, or improper PPE use. When it recognizes a potential violation, it immediately sends alerts to supervisors, allowing for rapid corrective action.

In a related development, **Hyundai Engineering & Construction** deployed AI-powered video analytics on its large-scale project sites to detect fatigue, unsafe posture, and even signs of microsleep. The AI system helped identify high-risk zones and times of day, leading to strategic rescheduling of work shifts and targeted safety briefings. A measurable reduction in near misses and minor injuries followed within months.

INTELLIGENT DATA AND PREDICTIVE ANALYTICS

The true promise of technology lies in not just responding to risk, but anticipating it. That's where AI in data analysis comes into play.

Shell, for example, uses machine learning algorithms to sift through volumes of incident reports, maintenance logs, and sensor data from its global operations. The goal? To predict where and when the next incident might happen. This predictive modeling allows safety leaders to deploy additional oversight, refresh training, or reassign personnel before issues arise. It's a shift from "fix after failure" to "prevent before it happens."

Similarly, a **peer-reviewed study by the Journal of Loss Prevention in the Process Industries** highlighted how the integration of AI in risk assessments led to a 25% improvement in the accuracy of identifying potential process safety hazards at a large petrochemical complex. Contractors benefited directly as risk profiles were better understood before work even began, leading to enhanced planning and reduced exposure.

IMMERSIVE LEARNING THROUGH VIRTUAL REALITY

Safety training is also experiencing a renaissance. Traditional classroom sessions are now being replaced or supplemented with **Virtual Reality (VR)** environments that place workers inside simulations of high-risk tasks. Imagine walking through a gas leak scenario on an offshore platform, or navigating the scaffolding of a high-rise under construction—all without leaving the training room.

At **ExxonMobil**, VR training modules have been introduced for both employees and contractors to safely rehearse emergency procedures. Workers can now "fail safely" in the simulation—learning from mistakes without real-world consequences. Early reports from the pilot program noted a significant increase in knowledge retention and faster response times during drills.

ENHANCING WELFARE AND MENTAL HEALTH MONITORING

Technology is also playing a role in **worker welfare**, particularly in addressing stress and mental health concerns—issues that are often silent but deeply disruptive. Today, wearables are being used to flag abnormal stress levels, erratic sleep patterns, or elevated heart rates over time. These insights, when anonymized and responsibly managed, allow wellness teams to intervene early.

For example, **Rio Tinto** piloted a mental wellness monitoring program using biometric data collected from wearables. Combined with AI analytics, the program identified correlations between certain work patterns and mental fatigue, allowing for revised shift structures and the provision of mental health support at key pressure points in the schedule.

REVIEW OF FAILURES: LESSONS WE HAVE LEARNED, OR MISTAKES WE WILL MAKE AGAIN?

Finally, let's talk about what happens when the project is done, everyone's been paid and you've started thinking about where you're going to be working next. Review the project that has just been completed. Give praise (publicly) to everyone involved in any success along the way. But the real purpose of this session is to establish what went wrong, what didn't go wrong but could have (near misses), and what lessons can be taken away from those experiences to improve performance in the future.

Too often in engineering and construction companies, anyone who looks back at these meetings a year or more later will see that they were really descriptions of mistakes that would be made again. And again. And, quite possibly, again after that.

Every project ends with an opportunity to improve. Take it. And accept my best wishes for a prosperous, successful, and SAFE future.

List of Book Citations

BIBLIOGRAPHY

1989 Phillips 66. 1989. *Phillips 66 Explosion in Pasadena, Texas.* U.S. Chemical Safety Board. Available at: https://www.csb.gov

2005 BP Texas City. 2007. *Investigation Report: Refinery Explosion and Fire.* U.S. Chemical Safety Board, Report No. 2005-04-I-TX. Available at: https://www.csb.gov

2009 Silver Eagle Refinery. 2010. *Investigation Report: Silver Eagle Refinery Hydrogen Pipeline Explosion.* U.S. Chemical Safety Board. Available at: https://www.csb.gov

2010 Deepwater Horizon. 2011. National Commission on the BP Deepwater Horizon Oil Spill and Offshore Drilling. *Deep Water: The Gulf Oil Disaster and the Future of Offshore Drilling.* Government Printing Office.

2010 Tesoro. 2014. *Final Report: Fatal Explosion and Fire at the Tesoro Anacortes Refinery.* U.S. Chemical Safety Board, Report No. 2010-08-I-WA. Available at: https://www.csb.gov

Airports Authority of India. 2022. *Terminal 2 Expansion: Health and Safety Report.* GVK Group. *[Internal report].*

Alcoa – O'Neill, Duhigg, Charles. 2012. *The Power of Habit: Why We Do What We Do in Life and Business.* New York: Random House.

Amazon Delivery Contractor Failures, Callahan, P., & O'Donovan, C. 2019. *Inside Amazon's Delivery System: Contractors Risk Death and Injury.* ProPublica/BuzzFeed News. Available at: https://www.propublica.org

Asiedu, R. & Adaku, E. 2019. "Cost Overruns of Public Sector Construction Projects: A Developing Country Perspective." *International Journal of Managing Projects in Business*, 13(1), 66–84.

BBC News. 2001. "Nike Faces New Sweatshop Claims." *BBC News.* Available at: http://news.bbc.co.uk

Bhopal Disaster (Historical Analysis) & Eckerman, I. 2005. *The Bhopal Saga: Causes and Consequences of the World's Largest Industrial Disaster.* Universities Press.

Bhopal Aftermath & Global Governance & Broughton, E. 2005. "The Bhopal Disaster and Its Aftermath: A Review." *Environmental Health*, 4(1), 6–11.

Brazil BYD Worker Welfare Incident & Phillips, T. 2024. "Brazil Shuts BYD Factory After Abuse of Chinese Workers." *The Guardian.* Available at: https://www.theguardian.com

Buncefield Explosion & Health and Safety Executive (HSE). 2011. *Buncefield: Why Did It Happen?* Final Report. Available at: https://www.hse.gov.uk

CBS News (Nike). 1996. *Nike Under Fire Over Sweatshops.* CBS Evening News. Available at: https://www.cbsnews.com

CCPS (Chevron Case Studies). Center for Chemical Process Safety. 2010. *Chevron Refinery Case Studies.* AIChE.

Dow/Union Carbide Governance & Fortun, K. 2001. *Advocacy After Bhopal: Environmentalism, Disaster, New Global Orders.* University of Chicago Press.

European Parliament, and Council of the European Union. 2012. *Directive 2012/18/EU of the European Parliament and of the Council of 4 July 2012 on the Control of Major-Accident Hazards Involving Dangerous Substances (Seveso III). Official Journal of the European Union* L 197 (24 July): 1–37

Fluor Corporation. 2023. *Zero Incident Performance (ZIP) Program.* Available at: https://www.fluor.com/sustainability/hse [Accessed 2 Aug. 2025].

Flyvbjerg, B. 2014. "What You Should Know About Megaprojects and Why." *Project Management Journal*, 45(2), 6–19.

Harvard Business Review—Trust Culture Link & Zak, P. J. 2017. "The Neuroscience of Trust." *Harvard Business Review,* Jan–Feb issue.

ILO—Fair Recruitment. International Labour Organization. 2014. *Fair Recruitment Initiative.* ILO. Available at: https://www.ilo.org

International Association of Oil & Gas Producers (IOGP). 2017. *Contractor HSE Management Guidelines*, Report 423. IOGP. Available at: https://www.iogp.org

Laing O'Rourke. 2023. *Workforce Wellbeing Strategy.* Available at: https://www.laingorourke.com [Accessed 2 Aug. 2025].

Locke, R. 2002. "The Promise and Peril of Globalization: The Case of Nike." *MIT Sloan Management Review*, 43(1), 77–86.

McDonald's—Labour Practices & Social Media & Guerra, C. 2021. "TikTokers Say McDonald's Is Broken." *Vox*. Available at: https://www.vox.com

Microsoft Cultural Transformation (Psychological Safety) & Nadella, S. 2017. *Hit Refresh: The Quest to Rediscover Microsoft's Soul.* Harper Business.

Nardello & Co. 2015. *Independent Review of Labor Practices at NYU Abu Dhabi.* Available at: https://www.nyu.edu

Petrofac. 2023. *Contractor Engagement Strategy—Middle East Operations. [Internal document].*

Petronas. 2021. *Contractor Health, Safety and Environment Management System (CHSEMS).* Available at: https://www.petronas.com

Qatar Labour Law. Government of Qatar. 2004. *Labour Law No. 14 of 2004.* Available at: https://www.ilo.org/dyn/natlex/docs/ELECTRONIC/67581/102549/F-1939266809/qat67581.pdf [Accessed 2 Aug. 2025].

Saudi Aramco. 2022. *Safety Management System (SMS) Manual.* Saudi Aramco.

Shell. 2023. *Contractor Safety Handbook.* Available at: https://www.shell.com/sustainability/safety/contractor-safety.html [Accessed 2 Aug. 2025].

The Guardian (Qatar Deaths), Pattisson, P., & McIntyre, N. 2021. "Revealed: 6,500 Migrant Workers Have Died in Qatar Since World Cup Award." *The Guardian.* Available at: https://www.theguardian.com/global-development/2021/feb/23/revealed-migrant-worker-deaths-qatar-fifa-world-cup [Accessed 2 Aug. 2025].

UN/ILO Multilateral Labour Governance. International Labour Office. 2020. *Global Labour Practices in Migrant Worker Supply Chains.* ILO.

Verité. 2020. *Forced Labor in Malaysia's Electronics Industry.* Verité. Available at: https://www.verite.org/research/electronics-malaysia/ [Accessed 2 Aug. 2025].

Index

H

I

J

K

L

M

W

Z